D0867380

An Introduction to the Philosophy of Science

AN INTRODUCTION TO THE PHILOSOPHY OF SCIENCE

Karel Lambert
Gordon G. Brittan, Jr.
University of California, Irvine

PRENTICE-HALL, INC.
ENGLEWOOD CLIFFS, N. J.

Printed in the United States of America.
Library of Congress
Catalog Card No.: 79-102288

AN INTRODUCTION TO THE PHILOSOPHY OF SCIENCE
Lambert and Brittan

C 13-492413-4
P 13-492405-3

PRENTICE-HALL INTERNATIONAL, INC., London
PRENTICE-HALL OF AUSTRALIA, PTY. LTD, Sydney
PRENTICE-HALL OF CANADA, LTD., Toronto
PRENTICE-HALL OF INDIA PRIVATE LIMITED, New Delhi
PRENTICE-HALL OF JAPAN, INC., Tokyo

Current printing (last number):
10 9 8 7 6 5 4 3 2

PREFACE

This book is intended for use in a quarter course in the philosophy of science at the sophomore-junior level. Three points of emphasis distinguish the present text from other available introductory texts.

The first emphasis is on arguments. A book of this kind, we believe, should above all else try to make clear the structure of arguments in the philosophy of science. Although it is introductory and presupposes virtually no background in philosophy, it is not a conversational rehash of opinions on the foundations of science; thus, it is especially well suited for students in a class situation.

The second emphasis is on the "natural dialectic" of these arguments. We have tried to set out the pattern of argument and counterargument in contemporary philosophy of science. Philosophy of science has been dominated in the last three decades by a group of philosophers, for example, Rudolf Carnap, H. Feigl, Carl G. Hempel, Ernest Nagel, and Hans Reichenbach, whose general viewpoint may be called the "classical" position. We begin by setting out the core of the classical position on the nature of mathematics, explanation and confirmation, and then discuss objections in each case. Most of these objections are of quite recent origin, some of them being our own, and as far as we know, they have not been gathered together in an introductory text in a systematic way. But a word of caution is in order. It would be wrong to assume that we are opposed to the classical position on the problems we discuss. On the contrary, to take one case as illustration, we are in considerable disagreement with each other on the topic of scientific laws, the senior author preferring the classical account and the junior author dissenting from same. We have tried to be fair and *neutral* on the issues we discuss. Any other policy, we feel, would be inappropriate in an introductory text.

The third emphasis is on the way in which the arguments are connected. There is a good deal of reference within each chapter to arguments in other chapters, and the final chapter, on the limits of scientific explanation, brings together the central strands in the arguments.

Introductory books in the philosophy of science tend to stress the natural sciences at the expense of, especially, the behavioral sciences. We have tried to redress the imbalance somewhat. The reader will find the amount of coverage given to topics of interest to the behavioral scientist (for example, to teleological and intentional explanation) in proportion to the total of the book to be considerably greater than is usually the case in introductory texts.

We have tried to indicate relevant bibliography—philosophical classics, standard accounts of the issues raised, more recent commentary in journal articles—in the rather extensive footnotes to each chapter. In addition, we have included a very short bibliography at the end of each chapter. By and large the items listed are more extensive, usually quite accessible, treatments of the topics taken up in the chapter. They are also generally more advanced. A comprehensive and up-to-date set of articles on many of the topics we treat is, Baruch Brody, *Readings in the Philosophy of Science* (Englewood Cliffs, N.J.: Prentice-Hall, Inc., 1970).

KAREL LAMBERT GORDON G. BRITTAN, JR.

ACKNOWLEDGMENTS

We wish to express our appreciation to those contemporary philosophers whose writings have literally determined the character of the discussion in this particular text. They are: C. D. Broad, Rudolf Carnap, Donald Davidson, Nelson Goodman, Carl G. Hempel, Jaakko Hintikka, T. S. Kuhn, Ernest Nagel, W. V. Quine, Bertrand Russell, Wesley Salmon, and Israel Scheffler. We are also grateful for the critical comments of our colleagues who saw the manuscript in various stages; but special thanks is due to Professors B. van Fraassen, D. Shapere, and S. Luchenbach.

K. L. G. G. B.

CONTENTS

An Introduction to the Philosophy of Science

INTRODUCTION

1

It is often said that the work philosophers do falls into two general categories. The first concerns individual areas of knowledge and experience; in our case the area of concern is science. The task of the philosopher here is to analyze conceptual and methodological issues; for example, are numbers to be taken as the primitive objects of mathematics, or are sets? Are explanation and prediction in science symmetrical? The next three chapters are concerned largely with issues of this type. But philosophers do another kind of work as well. Their second general task is to investigate how the various individual areas of knowledge and experience fit together, where their limits might lie, in order to gain some view of the whole. In the fifth and final chapter, we will turn to issues of this second type.

The two tasks are closely connected. For example, any attempt to determine the limits of scientific explanation depends on how the concept of scientific explanation is analysed. There is also a close connection between the analysis of the conceptual and methodological issues that arise in connection with a particular subject and the very characterization of that subject. The word "science" is used, perhaps now more than ever, in a variety of contexts with a variety of senses for a variety of purposes. We will not try to define "science"; but our discussion of certain central issues, chosen in part because they dominate the recent literature in philosophy of science, should help in the process of clarification. Science is so much with us that we all have a great many intuitions about what it involves. We shall try to sort out and examine a few of the more widely shared intuitions about the nature of science.

It is natural to begin a study of the foundation of science with remarks on the nature of mathematics. The traditional reason for doing so is the belief that science "presupposes" mathematics in the sense that it is an essentially quantitative subject. Our motives, however, are of a slightly different kind. We are not so much interested in the claim that mathematics is an indispensable tool of science as in the distinction that some philosophers and mathematicians have wanted to draw between them as "empirical" and "nonempirical" inquiries respectively. The distinction in question rests, of course, on a particular view of mathematics. One of the major aims of the next chapter is to sketch this view of mathematics (the "classical" view) with care, to set out the argument that has most often been given for it, and then to offer criticisms of it. An additional motive for discussing mathematics at the outset is that it introduces some technical vocabulary—"analytic," "synthetic," "logical truth," and so on—which will prove useful later.

The third chapter takes up the question of scientific explanation. While many philosophers emphasize the empirical character of science, others contend that what is distinctive about the scientific enterprise is the kind of explanations it affords. Once again our strategy is to elaborate the classical view of the matter and then to suggest criticisms of it. These criticisms fall very roughly into two groups. The first have to do with one aspect or another of the classical view: deductive versus inductive explanation, the nature of physical laws, and the alleged symmetry of explanation and prediction. These criticisms are mainly attempts to clarify the classical view of explanation and, in some cases, to see where it might have to be modified or amended. The second group of criticisms have to do with the claim, often advanced by those who share the classical view, that there is a single pattern of explanation in science. In particular, so the claim runs, there is no distinction to be made between the natural and social sciences as regards the *kind* of explanations exhibited. We examine three prima facie different patterns of explanation in examining this claim. Of special importance is the concept of teleological explanation and its role in the behavioral and social sciences. Whether a distinction can be made between the explanation of the behavior of physical objects and the explanation of the behavior of human beings, and more generally between the natural and behavioral or social sciences, depends in large part on whether teleological explanation is indispensable and/or irreducible.

The fourth chapter raises issues about confirmation and acceptability. These issues arise from the discussion in the previous two chapters in the following ways. On the one hand, the *empirical* character of science is said to stem from the fact that hypotheses in science, in contrast to other disciplines, for example, mathematics, are controlled by experimental test; they are confirmed or disconfirmed as the evidence for them is favorable or not. But the question is: what does it mean to say that a particular hypothesis is "confirmed by" or "falsified by" the facts? Indeed, it turns out that if we proceed to answer this question along certain apparently intuitive lines we are quickly led to paradox. Our discussion of confirmation concentrates on two of these "paradoxes of confirmation" and several of the resolutions of them that have been offered. On the other hand, questions concerning confirmation and acceptability also arise in the classical view of scientific explanation. For on that view, an event is explained by subsuming it under a law, and a natural way of characterizing laws is to say that they are general statements that are confirmed by their instances. But once again the question is, under what circumstances should we say that a statement is "confirmed by" its instances?

The final chapter draws on themes in the preceding three chapters. It addresses itself to the limits of scientific explanation. The arguments purporting to show that scientific explanation does have limits are directly related to the three patterns of explanation discussed in Chapter 3. Thus, the first is to the effect that human behavior cannot be scientifically explained—because human actions are voluntary—and thus concerns the concept of teleological explanation. The second argument contends that atomic theories of matter cannot, in principle, explain a number of natural phenomena and is directly concerned with the concept of reductive explanation. The third argument, that scientific explanations inevitably "leave something out" in their attempts to make the world comprehensible, is anchored in what we have so far called the classical view of scientific explanation.

In our discussion of these issues, which all in one way or another have to do with the empirical and explanatory character of science, we deal as well with other closely related and classical problems in the philosophy of science. Among these might be mentioned the analytic-synthetic distinction, the conventionalist theory of logical truth, the problem of induction, and the question whether the laws of nature are true or false.

THE NATURE OF MATHEMATICS

2

1. Introduction

What is mathematics? This is the first of four major problems we shall discuss in this book. Not only is the subject of great interest in its own right, but discussion of it should also help to further one of our eventual aims—to clarify what is meant by the term "science."

It is frequently claimed, for example, that mathematics is not genuinely a science (or, as is sometimes said, not genuinely an "empirical" science) because "it has no proper subject matter." Rather, the argument goes, its relation to science is more like that of form to content: mathematics is the language of science. Its usefulness lies in providing a) means for the precise expression of empirical hypotheses and b) methods for the difficult job of teasing out their consequences. The claim that mathematics is not genuinely a science is often linked to another. In contrast to statements of empirical science—for example, "The rate of response is a function of previous reinforcements"—strictly mathematical statements—for example, "The set of real numbers is nondenumerable"—are certain; no amount of evidence could ever serve to overturn them. Thus Albert Einstein's celebrated dictum: "As far as the propositions of mathematics refer to reality, they are not certain, and as far as they are certain they do not refer to reality."[1]

[1]Hermann Weyl, on page 134 of his book *Philosophy of Mathematics and Natural Science* (Princeton, N.J.: Princeton University Press, 1949), quotes this remark from Einstein's book, *Geometrie und Erfahrung.*

This view of the nature of mathematics has wide support among contemporary scientists, mathematicians, and philosophers. Why? For one thing, it allows scientists to avoid appeal to what might seem a ghostly realm of nonempirical, abstract "entities" (for example, numbers, points, and sets). For another, it allows mathematicians to distinguish between the work they do (so-called "pure" mathematics) and the applications to which that work is put by others. Finally, it allows empirically minded philosophers to account for the apparent certainty of mathematical statements, and at the same time, to deny that knowledge of nature is possible, independent of experience.

One representative of the view that mathematics has no subject matter is the mathematician-philosopher John Kemeny, who offers a particularly straightforward statement of it in his book, *A Philosopher Looks at Science.*[2] According to Kemeny, "Mathematics is a study of the form of arguments and . . . the most general branch of knowledge, but . . . it is completely devoid of subject matter" (p. 21). Moreover, "this irreplaceable language of Science, Mathematics, can never supply anything new" (p. 35). To believe otherwise, he maintains, is to confuse pure with applied mathematics; the latter does indeed have a subject matter that varies with the field of application, but the former, which is our present concern, does not.

How does Kemeny support this view? He bases it, in effect, on two major claims.

1: Mathematical truths reduce to (are nothing but) logical truths.
2: Logical truths are analytic, that is, true solely by virtue of the meanings of the words they contain.

It follows from these two claims that true "mathematical propositions are analytic" (p. 21). But if they are analytic, then mathematics is devoid of subject matter, although not in the sense that its statements are not "about" anything. The mathematical proposition

(i) $365 - 1 = 364$

which Kemeny takes as his example, is "about" numbers (if such there be); but the truth or falsity (the truthvalue) of (i), which is the concern of the mathematician, does not depend on what it is "about." The truth or falsity of such propositions "depends upon their form," by which Kemeny means "depends solely on how the

[2]From *A Philosopher Looks at Science* by J. G. Kemeny, Copyright © 1959, by Litton Educational Publishing, Inc., by permission of Van Nostrand Reinhold Co.

words they contain are used." To put it another way, the truthvalue of mathematical propositions is determined not by an inspection of nature but by an inspection of linguistic conventions.

For similar reasons, mathematical propositions are certain. "If we try to imagine . . . ways of testing the ordinary mathematical propositions, we find in each case that it is entirely irrelevant what the result of observation is, we *never* reject the propositions . . . Mathematical propositions are analytic a priori. They consist of an analysis of the meanings of words" (p. 18).

To get clear about Kemeny's view, it will be necessary to look more closely at the reduction of mathematical truths. Further, we shall have to examine what might be called "the linguistic theory" of logical truth. But first a better understanding of just which truths are logical truths is needed.

2. Logical Truths

Consider the statement "All men are men." Not only is the statement true, it remains true whatever noun or noun phrase we might substitute for "men" (in both of its occurrences). Even if we replace "men" by a noun that presumably does not refer to anything—say, "unicorns"—the statement still remains true. Statements that have this property we call *logical truths.*

This way of indicating the class of logical truths can be made somewhat more precise. If we divide the words appearing in a statement such as "All men are men" into two sorts, logical and descriptive—"All" and "are" are logical words, and "men" is a descriptive word—any statement that remains true under any and all substitutions of descriptive words or phrases for the descriptive words or phrases it contains is a logical truth. Alternatively, we could say that logical truths are true statements that contain only logical words *essentially.*[3] Clearly, "All men are men" is a logical truth.

We need to add three points of clarification. The first is that it is extremely difficult to give a precise standard for distinguishing between "logical" or "descriptive" expressions. But for our purposes this does not much matter; we can simply count those expressions that logicians class as "logical" as the logical expressions. In fact, four such expressions—"all," "or," "not," and "is-a-member-of"—suffice for the expression of the entire body of mathematics.[4]

[3]This rough criterion of logical truth, originally introduced by Bolzano, has been revived and considerably refined by W. V. Quine. See his *Mathematical Logic* (Cambridge, Mass.: Harvard University Press, 1951).

[4]Actually only three words are needed: "or" and "not" can be eliminated in favor of "neither-nor." The reasons for carrying out this discussion with the four mentioned above are pedagogical. In terms of our example, "are" is paraphrased

Second, our way of stating the criterion suggests that all logical truths contain both descriptive and logical words. But it is important to realize that there are many logical truths that contain no descriptive words. Here are two examples: "There is something that is identical with itself" and "Everything is identical with itself."[5] If these statements are true at all, they must be logical truths, for only logical words occur in them (and so occur in them essentially). Moreover, the existence of such examples is not an idle curiosity. Statements like "There exists something that is self-identical" will prove to be central later in this chapter.

The third point of clarification is that our criterion of logical truth does not commit us to the view that logical truths are true solely by virtue of the meanings of the words they contain and for this reason "certain." Indeed, our way of indicating the class of logical truths is perfectly consistent with the alternative views that logical truths delineate certain very general features of the world or that they express fundamental "laws of thought."

In other words, from the fact that we pick statements out as logical truths on the basis of linguistic characteristics—that is, that only logical words occur essentially in them—it does not follow that their truth or certainty derives solely from linguistic considerations. This conclusion requires a separate argument, which we shall consider when the linguistic theory of logical truth is taken up. Similarly, the fact that all the theorems of geometry are characterizable in a purely linguistic (or "syntactical") way as the set of statements deducible from a given list of axioms does not imply that the truth of these theorems can be traced to purely linguistic origins. For example, notice that plane geometry is often pictured as developing in the following manner. First, surveyors established the truth of a vast array of geometrical propositions on the basis of observations. Then Euclid systematized this collection by picking out a certain set of these true propositions as basic (the axioms) and deducing the remainder of the collection (the theorems) from this basic set. The point is that the *truth* of the theorems is not linguistic, though the *theoremhood* of these truths unquestionably is.

---

into the primitive logical vocabulary as follows: "everything either is a man or is not a man." Logical truths containing only "or" or "not" are sometimes called *tautologies*.

[5]"is-identical-with" is a logical expression, which can be defined in terms of "all," "or," "not," *and* "is-a-member-of." This is noteworthy, for it means that "is-identical-with" can be paraphrased only with the help of "is-a-member-of," a term of set theory. There are reasons, to which we shall turn shortly, for not treating set theory as a branch of "logic."

3. The Reduction of Mathematics

The first of the two claims on which the view of mathematics we are considering is based is that mathematical truths reduce to (are nothing but) logical truths. Typically, the claim is supported by showing that all mathematical statements are deducible, with only the help of logical rules, from a handful of logical principles. To show this is to show that any ostensibly mathematical statement can be paraphrased in such a way that the only words that occur in it essentially are logical words. Mathematics reduces, in this sense, to logic. The attempt to show this—a program known as *logicism* and associated with the names of Frege, Russell, and Whitehead—is one of the great intellectual adventures of modern times. The logicist program is set forth in greatest detail in Russell and Whitehead's *Principia Mathematica*.[6] Here we must content ourselves with the barest sketch of it.

The first step in the program is to reduce the various branches of mathematics, for example, analysis, algebra, geometry, to arithmetic, on the model of Descartes' reduction of geometry to algebra via analytic geometry. As Kemeny puts it: "It can be shown that all of Mathematics is founded on properties of integers (whole numbers). If you are well acquainted with these, the rest of Mathematics is deducible by purely logical arguments. So in a sense the nature of Mathematics can be identified with that of the theory of integers" (p. 20).

The second step is to show how arithmetic, "the theory of integers," can be reduced to logic. In *Principia*, this is accomplished somewhat as follows: we take it that arithmetic can be developed on the basis of five axioms set down late in the last century by the Italian mathematician Peano; from them all the properties of the integers can be derived by strictly logical reasoning.[7] The five axioms are:

A.1: 0 is a number.
A.2: The successor of any number is a number.
A.3: No two numbers have the same successor.
A.4: 0 is not the successor of any number.
A.5: If P is a predicate true of 0, and if whenever P is true of a number n, it is also true of the successor of n, then P is true of every number.

[6]Gottlob Frege, *The Foundations of Arithmetic*, trans. J. L. Austin (New York: Harper and Row, Publishers, 1953), is an excellent introduction to the logicist program.

[7]Actually, these five axioms suffice for only a fragment of arithmetic, but the point in question is not affected.

The trick is to define the arithmetical notions that appear in these axioms in logical terms. Once this is done, the reduction of mathematics to logic proceeds without difficulty. Once numbers are construed in logical terms, the basic arithmetical operations (addition and multiplication) can be similarly defined. In this way, mathematical statements turn out to be *provable as theorems in logic.*

The three notions involved are "0," "is a number," and "is the successor of." In fact, the second can be defined in terms of the first and the third. To say that n is a natural number is to say that n is 0 or is the successor of 0 or is the successor of the successor of 0, and so on.[8] We have only to deal with "0" and "is the successor of." "0" can be defined as the set that contains only the set that has no members—that is, the set that contains the null set as its only member. And the successor of any number n is the set of all sets that, when deprived of a member, come to belong to n. Since these defining expressions in turn can be analyzed in terms of "all," "or," "not," and "is-a-member-of," which are logical expressions, the reduction of mathematics to logic is virtually complete. To quote Kemeny once more, Russell and Whitehead "show that the mathematical concepts used by Peano can be defined in terms of logical words and that all their properties can be demonstrated by pure logic. Thus Mathematics is shown to be no more than highly developed Logic" (p. 21).[9]

4. The Linguistic Theory of Truth

The second claim on which Kemeny's argument turns is that logical truths are analytic, that is, true by virtue of the meanings of the words they contain. As we have already mentioned, our characterization of the class of logical truths does not commit us to any special doctrine about where their truth comes from, nor does it enable us to infer that logical truths are "certain." The linguistic theory of logical truth is such a special doctrine, and it does allow us to draw the intended conclusions. There are other answers—for example that logical truths are certain because they describe the most general and pervasive traits of reality—but, as

[8]One of Frege's principal contributions was to analyze out the "etc." On his version, n is a natural number just in case n is a member of every set X such that 0 is a member of X and all successors of member of X are members of X.

[9]Kemeny goes on to add the following crucial proviso: "In this process two new logical principles turn up, the axioms of infinity and choice, whose somewhat controversial nature need not concern us here. Let it suffice that if we recognize these two as legitimate logical principles—as most logicians do—then all of Mathematics follows and becomes just advanced Logic."

we have also had occasion to mention, an answer of this kind does not recommend itself to the empirically-minded philosopher.[10]

Proponents of the linguistic theory say that we are to regard logical truths as a limiting case in the class of true statements. Many true statements have both a factual and a linguistic component. Thus, the truth of "The grass is green" depends both on what I mean by "grass," "green," and so on, and on whether or not the grass is in fact green. The truth of the statement "The grass is green or it is not green," however, seems to depend solely on how the words that it contains are used; that is, the truth of the statement in question depends solely on the meanings of the logical words "is," "or," and "not," and on the understanding that "green," in both of its occurrences, has the same meaning (whatever it may be). In the same way, logical truths generally owe their truth to linguistic, not factual, sources; their truth is determined solely by the ways in which the words they contain are used. For example, if "The grass is green or it is not green" were false, it would be because the words it contains had different meanings than the ones we normally associate with them, not because the world had been misdescribed. This theory about logical truth is sometimes loosely stated by saying that logical (and ultimately mathematical) truths are "true by definition." Since they are "true by definition," moreover, such truths cannot be overturned by empirical evidence; they are certain.

The linguistic theory is also sometimes stated by saying that logical truths are "true by convention." In the sense that all definitions are conventional, of course, this variant of the theory is closely tied to the one discussed in the preceding paragraph. What it comes to is this: on our criterion, logical truths are those true statements that contain only logical words essentially; but, the argument goes, the meaning of these logical words "or," "not," and so on, is fixed conventionally. For example, logicians decide that statements compounded with "or" shall be true just in case at least one of the components is true. It is because the meanings of the logical words have been fixed in certain ways that the sentences in which they alone occur essentially are true. Conversely, if we had adopted a different set of conventions concerning their use, the class of truths in which they alone figure essentially would not be the same. But, once again, if logical truths are "true by convention," then it seems

[10]Some of what follows derives from W. V. Quine's essay, "Carnap and Logical Truth," which is in *The Ways of Paradox* (New York: Random House, Inc., 1966).

to follow that they must be certain, that their truth is independent of how matters stand in the world.

Taken together with the first claim, that mathematical truths reduce to logical truths, the linguistic theory extends to mathematical truths as well. Thus Kemeny: a (true) mathematical proposition is "true because of the meanings of the terms; it is a true analytic proposition" (pp. 21–22). We know that, for example, "365 — 1 = 364" is true when we know what "365," "— 1," "=," and "364" mean. But knowing what these mathematical expressions mean comes, if mathematics reduces to logic, to knowing what the logical expressions in terms of which they can be defined mean. But we do know what these expressions mean; their meanings have been determined in accordance with certain usages. Similarly, the truth of the theorems of Euclidean geometry, for example, the theorem "A straight line is the shortest distance between two points," is determined by the fact that "point," "line," "plane," and so on have the meanings they do, and not, for example, by the (alleged) fact that the physical world is Euclidean in structure. This, to repeat a point made earlier, would be to confuse pure with applied mathematics. Statements of applied mathematics are not true solely by virtue of the meanings of the constituent words; that is, they are not analytic, but neither are they "certain."

A final point. The above claims are not unconnected. How logical truth is characterized and how the reduction at stake is to be appraised both turn in large part on what we understand by "logic." Conclusions will vary, depending on the latitude we allow ourselves regarding this term.

5. Logic and Mathematics

Notice that the reduction of mathematics, by way of defining "0" and "successor" in logical terms, involves the notion of a *set*.[11] If mathematics is reducible to logic, then "logic" must be understood to include set theory. There are reasons, however, for distinguishing between logic in a narrower sense and set theory, and ultimately for distinguishing between mathematics and logic.

For example, if we were to look more carefully at Russell and Whitehead's program in *Principia*, we would see that it depended essentially on the use of two axioms—the axiom of infinity and the axiom of choice—that are extremely difficult to construe as principles whose truth or falsity depends solely on the meanings of

[11]Hence, it involves the expression "is-a-member-of." This expression cannot be adequately analyzed only in terms of "all," "or," and "not"; set theory cannot be reduced to a logic that talks only about individuals.

the constituent words. That there exists an infinite number of individuals, which is what the axiom of infinity asserts, seems to be true neither by virtue of the meanings of the words "exists" and "infinite" nor by virtue solely of the meanings of the logical words involved. Scrutinizing the statement is of little help. The axiom is true, if at all, just in case it is a fact that there exist an infinite number of individuals. Similarly, the axiom of choice requires more than an imaginative stretch to be regarded by one as a principle of logic.[12]

This point is worth developing. Both of these controversial axioms make existence claims. In fact, set theory generally contains a variety of assertions concerning the existence of certain sorts of things. Consider the statement that there exists a null set, a set with no members. This statement (which contains no descriptive words, by the way) certainly appears to be about something—namely, the null set. If those who contend that mathematics is not really about anything, because the truth of the statements with which it is concerned stems from the arbitrarily assigned meanings of the words making them up, want to establish their claim, they must argue that the reliance of the truth of such statements on what they are about is merely apparent; a way of speaking at best. But suppose there were no objects at all, or at least no sets. Then the statement that *there exists* a null set could not be true. Whether there is anything, whether something exists depends on the facts and is certainly not solely dependent on how we use words. That neutrinos, or habits, or even God (*pace* St. Anselm) exist(s) is not decidable by appeal to the meanings of the words "neutrino," "habit," "God," and "exist(s)." However these statements are to be established, their truth or falsity does not come from the use of words alone. The claim that the truth of, for example, "There exists a null set" only *appears* to depend on what it is about is a claim that itself rests on an illusion.

An objection that thus arises immediately to Kemeny's view of the nature of mathematics is simply that many mathematical truths—indeed, many ostensibly "logical" truths (witness "There exists something that is self-identical")—are not analytic. Thus, insofar as these statements depend upon the "facts" for their truth,

[12]The axiom of choice asserts that there is a way of picking a single member from each of a list of disjoint sets such that the selected members comprise a new set distinct from all others in the original list. The truth of this axiom, if it is true, requires that there *be* such a way (that is, such a function); it is not enough that the words involved in its expression have the meanings they do.

mathematics does have a subject matter and what it is about—sets, numbers, or whatever—is an important area of investigation. This objection is directed more against a certain theory of logical truth—we have called it the linguistic theory—than against the claim that mathematics reduces to logic. Yet the above remarks bear on this last claim also.

According to Leibniz, the seventeenth century mathematician and philosopher, a logical truth is one that is true in all possible "worlds." The notion of a possible world is a difficult one; contemporary philosophers of logic tend to visualize a possible world as a set of objects with their properties. But so long as they do so, one possible world is the world that contains nothing. In recent years, logicians of a philosophical bent have constructed logical systems, all of whose theorems hold in all possible worlds *including the empty one.*[13] Now this reformulation of logic in accordance with Leibniz' characterization of logical truth as truth in *all* possible worlds has the effect of eliminating statements like "There exists something that is self-identical" and, indeed, any statement that begins with "There exists a . . ." as truths of logic. The statement in question would only be true in "worlds" having at least one member, but would be false in the empty world. Since the Quine-Bolzano standard of logical truth does not exclude such statements from the class of logical truths, it may reasonably be claimed not to be an equivalent formulation of Leibniz' characterization so long as possible worlds are construed as *sets* of objects.[14] There is, however, a deeper point. Logic reformulated in the above way contains no existence claims. Since set theory abounds in them, it is not the case that mathematics reduces to logic, because the alleged reduction takes place, so the argument goes, via set theory.[15]

Whether logic contains existence claims or not, there is another argument often advanced to show that mathematical truth does not

[13]See Karel Lambert's paper, "Free Logic and the Concept of Existence," in the *Notre Dame Journal of Formal Logic,* vol. VIII (April, 1967), pp. 135–41, for an example of such a system.

[14]See the paper "Logical Truth Revisited" by P. Hinman, J. Kim, and S. Stitch in the *Journal of Philosophy,* LXV, (September, 1968), 495–500, for a good discussion of the Quine-Bolzano standard. See also D. Berlinski and D. Gallin, "Quine's Definition of Logical Truth," *Noûs,* III, No. 2, (1969), 111–28.

[15]Earlier in this chapter, Frege was mentioned as one of the founders of Logicism, the position now being criticized. Apparently, in his later years, Frege rejected Logicism on grounds similar to those presented in this, and in the preceding paragraphs of this section. See P. Benacerraf and H. Putnam, eds., *Philosophy of Mathematics: Selected Readings* (Englewood Cliffs, N.J.: Prentice-Hall, Inc., 1964), p. 11.

reduce to logical truth, an argument based on a celebrated discovery by the mathematical logician Kurt Gödel.[16] According to Gödel's "incompleteness theorem," there are mathematical truths that are not provable within the resources of a formal system designed to yield such proofs. In short, mathematical truth does not coincide with proof in a formal system. The point is that logic can be formalized—that is, it can be organized in such a way that its theorems (provable formulas) can be got in a finite number of steps from certain explicit axioms by certain syntactical rules. Further, it can be shown that the class of theorems coincides with the class of logical truths, a result also first proved by Gödel. Accordingly, if mathematical truth reduced (by derivation) to logical truth, mathematical truth would coincide with proof in a formal system, namely, formalized logic. But this is just what Gödel's "incompleteness theorem" precludes. Hence, mathematics cannot reduce to logic so long, once again, as logic is thought of as not including set theory. This objection presents us with a dilemma: either logic does not include set theory, or the class of logical truths (understood to include the truths of set theory) does not coincide with provability in a formal system. If one is prepared to accept the unprovability of a whole class of logical truths, then one need not boggle, on that account, at the inclusion of set theory in logic.

6. Truth by Convention

Kemeny's view about the conventional character of mathematical truth is also open to objection. Specifically, it might be urged, he seems to misplace what really is conventional and arbitrary in logical and mathematical statements. Consider once again the statement that there exists a null set. In most versions of set theory, this statement, if not an axiom itself, is deducible from the axiom that says: for every property there exists a set consisting of all and only those objects that have that property.[17] So the statement that there exists an empty set has a subject matter provided the axiom from which it is derived has a subject matter. It is at this point that the misplacement enters. It is *not* the truth of the theorem that is arbitrary or conventional, but rather the particular truth that we choose to be an axiom. To put it another way, *axiom-*

[16]There is a clear and useful account of Gödel's discovery in Ernest Nagel and James R. Newman, *Gödel's Proof* (New York: New York University Press, 1960).

[17]It should be remarked that unless restricted this axiom leads straight to paradox—that is, Russell's. But the needed restrictions do not bear on the issue at hand.

hood is conventional, *truth* is not. So much is hidden by the expression "true by convention": the expression might mean *either* that the truth of a statement is conventional *or* that it is a particular true statement chosen to be an axiom that is conventional. But even a cursory examination of the history of mathematics suffices to show that what is arbitrary in the behavior of mathematicians has to do with which truth (or truths) in a batch of truths is taken as a sufficient basis for producing the rest of the statements in a given branch of mathematics. Euclid's geometry is a prime example. This objection to the "truth by convention" thesis is suggested by W. V. Quine.[18]

Another objection is that the linguistic theory of logical truth is vitiated by a subtle ambiguity. Recall that the theory asserts that the truth of a truth of logic is *determined* solely by the meanings of the constituent words. Equivalent statements of the theory may be obtained by replacing "determined" by "depends on" or "in virtue of" (and rephrasing the rest of the statement, if need be, to make grammatical sense). The point is that the linguistic theory seems to be unavoidable when one examines the methods logicians use in determining logical truth. For example, let us look again at propositional logic. A classical method used to determine (propositional) logical truth is the so-called tabular method. Very roughly, it can be described as a method for computing the truth of compound statements on the basis of the truth values of their simple components. Thus, suppose that one takes the statement "John is tall" as the only simple (or *atomic*) statement in a language where the means of constructing compounds are the conjunctions "not" and "or." (An atomic statement is one which contains neither "not" nor "or.") Now we may lay down "conventions" for computing the truth and eventually the logical truth of compounds like "John is not tall" or "John is tall or John is not tall," and so on. The "conventions" are essentially threefold:[19]

1. Assign every simple statement either the value "true" or the value "false" (on the assumption that every statement is either true or false).
2. "Define" the logical conjunctions "or" and "not" in such a way that one can tell the truthvalue of a statement containing these conjunctions (for example, we can "define" a statement of the

[18]W. V. Quine, "Carnap and Logical Truth," in *The Ways of Paradox*.

[19]The present conventions are thus conventions that allow us to evaluate the truth of any statement, simple or compound, in which the logical words may occur either essentially or inessentially.

form "——— is not . . . ." as true when "——— is . . . " is false, and vice versa; similarly, a statement of the form "——— or . . . . " is true just in case at least one of "———" or " . . . . " is true, otherwise false.

3. Compute the truthvalues of the compounds on the basis of 1. and 2. (for example, if we know the truthvalue of "John is tall"—say it is true—then "John is not tall" must be false, given our definition of "——— is not . . . . ").

Now certain statements will turn out true by this method no matter what the truthvalues of their component atomic statements. The statement "John is tall or he is not tall" is such a statement, because if "John is tall" is true, the compound is true, and if "John is tall" is false, the compound is still true.

Note that the statement in question is also logically true by either the Quine-Bolzano or the Leibniz ("possible worlds") standard. Accordingly, the tabular method enables us to pick out a large class of logical truths. Now as we mentioned above, it seems perfectly appropriate to say that statements like "John is tall or he is not tall" have their truth determined solely by appeal to the meanings of their constituent (logical) words—in this case by appeal to the meanings of the words "not" and "or." The linguistic theory interprets this perfectly appropriate claim as a statement about the *source* of logical truth. But it is just as plausible to say that what the tabular method does is to "determine" logical truths merely in the sense of picking them out, and that the truth of logical truths does not come from the linguistic "conventions" (rules and definitions) used to evaluate compound statements. The same may be said for the linguistic theory of logical truth expressed in terms of "depends on" or "in virtue of." In other words, the "conventions" associated with methods for ascertaining logical truth can be thought of as *devices for identifying* (picking out) logical truths, in the way that fingerprints are used to identify men, and not as the *sources* of the truth of statements such as "John is tall or he is not tall." The objection, then, is that the linguistic theory of logical truth is at best unproved, and that its force is considerably diminished when the ambiguity of "determined" and its synonyms is noted.

7. Mathematics and Science

The argument we have been considering has two main premises —one about the assimilation of mathematics to logic and the other about the linguistic theory of logical truth. Neither, as we have seen, is uncontroversial. But that is not the end of the matter; for it still might be felt that, however unconvincing the argument, the intuitive contrast it was intended to support—between scientific and mathematical statements—is real. Mathematical statements differ sharply from scientific statements in that only the former are *certain* and that *empirical evidence is irrelevant* to their truth or falsity. In this section we shall discuss very briefly alternative accounts that many have held to support the contrast successfully.

The first of these touches on a point that we have already made. We suggested that, in one sense, mathematics does have a subject matter and is not to be distinguished in this respect from science, but that its subject matter—sets for one thing—seems to be different. To put it in an almost question-begging way, mathematics has a nonempirical subject matter, whereas the subject matter that is the concern of scientists consists of observable, spatially and temporally locatable objects and events. Mathematics seems to be about objects that are unobservable and that have no spatio-temporal location. This is what is usually meant, for instance, by the assertion that sets are *abstract* objects.

Of course, if anyone were to show that mathematics could dispense with abstract objects—that in the case at hand one could eliminate sets in favor of spatio-temporal objects then the alleged contrast with science would collapse.[20] But there are more immediately available reasons for rejecting the present contrast between mathematics and empirical science. On the one hand, it is at least very controversial whether observable, spatio-temporal objects and events exhaust the subject matter of science (properly so-called). Many of the objects, (for example, submicroscopic particles) that science studies are not in any straightforward sense "observable," and in the more theoretical reaches of science many qualifications have to be made about "spatio-temporal locatability." On the other hand, it has been claimed[21] that, while they are perhaps not spatio-temporal, mathematical objects are in some sense "observable." Moreover, it seems difficult if not impossible to contrast these senses of "observable" without begging the question.[22]

[20]Interestingly enough, Russell made the attempt to get along without sets in *Principia* (on the basis of a so-called "no-class" theory that appeared to be committed to the existence of individuals only). He failed, however, to do away with abstract objects. His elimination succeeds only at the expense of introducing *attributes*, which are just another variety of abstract object.

[21]Most notably by Kurt Gödel. See his paper, "Cantor's Continuum Problem," first published in 1947 and reprinted in P. Benacerraf and H. Putnam, eds., *Philosophy of Mathematics: Selected Readings* (Englewood Cliffs, N.J.: Prentice-Hall, Inc., 1964).

[22]See N. R. Hanson, *Patterns of Discovery* (London: Cambridge University Press, 1958), Chap. 1. The whole notion of "observability" is, despite much discussion, very unclear.

There is a related way of drawing a line between science and mathematics—not in terms of their subject matter, but in terms of the *methods* used to demonstrate the truth and falsity of statements. Mathematical statements are shown to be true or false only by analytical procedures, for example, like deduction; empirical statements in contrast require observation to demonstrate their truth or falsity.

A moment's reflection, however, should make it clear that distinguishing between mathematical and scientific statements on the basis of the methods used to establish them will not work. Many of the "higher order" statements in physics, for example, are not directly testable by observation, nor are immediate consequences of these statements. Often such statements are justified by the simple fact that they are consequences of other, already accepted statements. Indeed, we can go so far as to say that it is doubtful whether a "higher order" statement would be accepted purely on the basis of empirical evidence.[23] But if the acceptability of many statements of science is simply a consequence of their implying or being implied by other already accepted statements of a theory, then there is no contrast with mathematics; for whether a statement implies or is implied by another is determined by strictly analytical procedures.

There remain those who protest that, although it is perhaps only very indirect, there is a connection between empirical matters of fact and scientific statements that does not obtain in the case of mathematical statements. Somewhere in the hierarchy of statements that make up a scientific theory there is contact with the world, so that at least some of the statements have observational consequences; but the same is not true of any system of pure (that is, not applied) mathematical statements. This claim has to do more with the statements of mathematics and science considered collectively than individually, and in this respect it differs from the arguments already considered. There seem to be at least two ways in which it can be interpreted.

One way proceeds in terms of a notion of "factual enrichment." Consider the following three statements:

1. John is tall.
2. Henry is fat.
3. The rat ran right.

We label these *atomic* (or *simple*) statements because they are not compounded of other statements using the logical conjunctions "or"

[23]Deductive connections with other statements in a theory is important to determining whether a statement alleged to be a law really is. This matter will be discussed in the next chapter.

and "not," and we can also take them to be *observation* statements, in the sense that their truth or falsity can more or less be determined by observation. Now an interesting fact about logic can be noted: it is not possible to enlarge this list of the three atomic statements by means of the logical words alone and the rules governing their use. Logic can tell us how to compound these statements (in truth-preserving ways), but it cannot add other atomic statements to our basic stock. This is what one means when one says that logic is merely a set of rules for transforming statements into other statements, and for this reason is not factually enriching.

Furthermore, even if we add mathematics to logic, we cannot get any additional simple observation statements than the three we began with. The addition of mathematics does not increase our ability to get new factual information of this type. Thus mathematics too is factually unenriching. The proof of this claim is not difficult.[24] Briefly, it can be shown a) that three axioms of set theory —the so-called axioms of comprehension (suitably restricted), extensionality, and choice—added to the principles of logic suffice for the development of mathematics, and b) that the addition of these three axioms to logic does not permit us to derive any more atomic observation statements than we could before they were added. In view of their inability to increase our factual knowledge in this sense, mathematical statements may be said to be contrasted with at least some of the more or less "observational" statements that make up a given scientific theory.

From this point of view, and not from the questionable thesis that mathematics has no subject matter, there may be something to the claim that mathematics is the language of science. It provides the means whereby facts can be expressed and the relations between them made clear. If mathematics alone cannot lead us to new, simple observation statements, it can greatly enrich the ways in which these statements can be expressed and combined.

At the same time, this argument does not appear to take us very far toward supporting the contrast between mathematical and scientific statements in terms of certainty. It gives us no reason to suppose that mathematical statements are any more or less certain than scientific statements.

The second way to interpret the claim that mathematical statements do not stand in the same relation to experience as empirical scientific statements is that the former cannot be refuted by the facts. They are true, if at all, come what may; empirical

[24]Those interested can find it in an article by Hilary Putnam. "Mathematics and the Existence of Abstract Entities," *Philosophical Studies*, VII (1956), 81–88.

scientific statements, on the other hand, are always subject to revision in the face of recalcitrant experience in the laboratory, and so are not certain or necessary. At this point, we make contact again with Kemeny's view of the nature of mathematics. Mathematical statements for Kemeny—and for those who share his position—are true or false independent of how matters stand in the world; they are *analytic*. Empirical scientific statements, on the contrary, are *synthetic*: they are not true because of their form, but because of their content; that is, because of the way in which they are confirmed by observation, however indirectly.

This view is very widely accepted as true, although it should be noted at once that as it stands it is not very precise. We have already had something to say about the lack of clarity in "observation." "Confirmed" will be taken up in the fourth chapter. Switching to an alternative vocabulary and saying that mathematical, in contrast to scientific, statements are *necessary* is not very helpful either, for in the attempt to make *this* expression clear, we are quickly led back to talk about "confirmed" and "observation" (for example, a statement is necessary just in case there is no possibility of its being disconfirmed by observation, and so on).

There are powerful objections to the view that there is a sharp distinction between mathematical and scientific statements on the basis that the former are analytic and the latter synthetic. Heretical as it might sound, a case can be made for saying that mathematical statements too are "always subject to revision in the face of recalcitrant experience in the laboratory."

This case can best be appreciated by way of examples. We shall focus on two. Consider the statement that momentum is proportional to velocity as one paradigm of an analytic statement.[25] "Momentum" is here simply *defined* as "mass times velocity"—that is what "momentum" *means*. Now suppose this statement to be part of a scientific theory (for example, classical physics) with which certain experimental findings (for example, the Michelson-Morley experiment) conflict. Since these findings do not conflict with any one statement of the theory in particular, there are a variety of ways in which the theory might be revised in order to accommodate them. One such revision that serves to align theory with data

[25]The example is W. V. Quine's and what follows draws heavily on his article "Necessary Truth," first published in 1963 and reprinted in *The Ways of Paradox*. See also H. Putnam, "The Analytic and the Synthetic," in *Minnesota Studies in the Philosophy of Science*, Vol. III, eds. H. Feigl and G. Maxwell (Minneapolis, Minn.: University of Minnesota Press, 1962), pp. 358–97.

amounts to amending the statement that momentum is proportional to velocity. We might say (as a matter of definition, or, better yet, redefinition) that momentum is inversely proportional to the square root of one minus the velocity squared over the speed of light. This apparently was Einstein's course. It allowed him to accommodate the Michelson-Morley experiment, among others, into a physics that was in many other respects "classical." But if an "analytic statement," as the one in question ostensibly is, can be revised in this way when the theory in which it belongs is confronted by recalcitrant experience, there seems to be no reason for contrasting it with other, ostensibly "synthetic" statements, which the theory also contains.

An even more striking case concerns contemporary developments in physics, in particular quantum mechanics. This time our paradigmatic analytic statement is the law of excluded middle, "*p* or not *p*." This is a fundamental logical principle, if there are any. It meets every criterion one might want to propose for an analytic statement—logical truth, necessary proposition, and so on. But according to Heisenberg's "Uncertainty Principle," a simultaneous determination of a particle's momentum and position is impossible. It is, for example, not possible (on that principle) to maintain that an object has, at a given moment, such and such momentum and position, or that it does not. As a result, certain philosophers[26] have recommended abandoning the law of excluded middle and choosing instead a logic in which it does not figure as a logical truth, which is, once again, to revise even one's logic in the face of recalcitrant experience.

These two cases suggest that although at any given time, within any particular scientific theory, certain statements are not held liable to revision, there is no reason why they should not be. Thus, if certain predictions in a discipline such as mechanics fail, one does not typically embark on a revision of the differential calculus, though the calculus is part of that theory. But there are circumstances in which one might conceivably do so. Statements that are, relative to certain contexts, immune to revision in the face of the facts might be called "analytic" or "certain." The important point is that whether such statements are construed as "analytic" depends on the context. A statement can be taken as definitional in one

[26]Notably Hans Reichenbach, in his book *Philosophic Foundations of Quantum Mechanics* (Berkeley: University of California Press, 1944). Other philosophers (those of the so-called "intuitionist" school) have recommended abandonment of the same law in mathematics, on the basis of quite different considerations.

interpretation of a theory, as a natural law in another interpretation, and so on. Newton's second law— $F = ma$ (force equals mass times acceleration)—has been variously interpreted in just these ways. The truth or falsity of the statement is not simply determined either by its form or by its content, but rather by the role it plays in those theories in which it figures. Thus one can, relative to certain theories and contexts (for example, when the theory is being tested), make a distinction between "analytic" and "synthetic" statements. The "analytic" statements will be those held *constant*;[27] recalcitrant experience will have no bearing on their truth or falsity. In most cases the statements held constant will be the logical and mathematical ones.[28] But there are cases like those sketched, in which even logical and mathematical principles may be subject to revision. Mathematical statements do not typically stand in the same close relationships to experience that empirical statements do, and they are to that extent "more necessary," less subject to revision—but the difference seems to be one of degree only. Any sharp boundary between mathematics and science is, perhaps, no more than terminological.

For Further Reading

Barker, Stephen, *The Philosophy of Mathematics*. Englewood Cliffs, N.J.: Prentice-Hall, Inc., 1964.

Benacerraf, P., and H. Putnam, eds., *Philosophy of Mathematics: Selected Readings*. Englewood Cliffs, N.J.: Prentice-Hall, Inc., 1964.

Frege, Gottlob, *The Foundations of Arithmetic*, trans., J. L. Austin. New York: Harper & Row, Publishers, 1953.

Kemeny, John, *A Philosopher Looks at Science*. New York: Van Nostrand Reinhold Co., 1959.

[27]An interesting characterization of "analytic" in this spirit may be found in Bas van Fraassen's paper, "Meaning Relations Among Predicates" in *Noûs*, I (1967), 161–81.

[28]One reason we are so reluctant to revise them is because of their very great generality.

Korner, Stephan, *The Philosophy of Mathematics.* New York: Harper & Row, Publishers, 1960.

Quine, W. V., *The Ways of Paradox.* New York: Random House, Inc., 1966.

Russell, Bertrand, *Introduction to Mathematical Philosophy.* London: George Allen and Unwin Ltd., 1918.

EXPLANATION

3

1. Introduction

". . . the distinctive aim of the scientific enterprise," a well known philosopher of science writes, "is to provide systematic and responsibly supported explanations."[1] Few would disagree. The rub comes in trying to make clear a) what counts as "responsible support" and b) how the concept of explanation is to be analyzed. We will discuss the latter question in this chapter and the former in the next.

These are interesting and difficult questions. An important source of their difficulty lies in the fact that words like "explanation" have no single clear and distinct sense. We all have a number of intuitions about what it is to give an explanation, many of them vague, perhaps some of them inconsistent. Indeed, we often find it difficult to say whether a particular account of some phenomenon qualifies as an explanation of it or not. Moreover, although roughly speaking, an explanation is an answer to a "why" question, there are many different sorts of "why" questions and a variety of ways in which they can be answered. Consider a case in point.[2] One migh in asking "Why did Descartes believe in God?" be asking for *reason* in support of his belief, a justification or defense of it. An appro

[1]Ernest Nagel, *The Structure of Science* (New York: Harcourt, Brace & Worl Inc., 1961), p. 15.

[2]The example is Morton White's, *The Foundations of Historical Knowledg* (New York: Harper & Row, Publishers, 1965).

priate answer would be to cite the "ontological argument" that he gives in the fifth *Meditation*. Or we might in asking "Why did Descartes believe in God?" be asking for the causes of this belief. An appropriate (and frequently given) answer this time would be to cite Descartes' Catholic upbringing. Of course, the reasons for his belief might also have been the causes of it, although many people today, having read Marx and Freud,[3] would perhaps regard this possibility as remote. But they need not, and initially it seems doubtful whether a single concept of explanation covers both sorts of cases.

We shall not look, then, for an analysis of the concept of explanation that covers every sort of "why" question or that captures all our intuitions on the subject. Rather, we shall begin by fixing on one sort of "why" question and taking only some of our intuitions as central, dismissing others as relatively unimportant. In every case of analyzing a concept, there is some legislating; otherwise the attempt to clarify and to be precise would have to be given up.

Recall, for example, the analysis of the concept of number advanced by Frege, Russell, and Whitehead. Their analysis certainly does not cover every use of the word "number" nor catch intuitions some people have about numbers.[4] But it does serve to incorporate many of these uses and intuitions. In particular, it is widely agreed that an important requirement on the adequacy of any analysis of the concept of number is that it satisfy the Peano Axioms for arithmetic listed in the second chapter. These axioms characterize certain very basic properties of numbers. The set-theoretical construction that Frege, Russell, and Whitehead propose meets this requirement; to this extent it constitutes an adequate analysis.[5]

Guided by these remarks, let us begin by restricting our attention to so-called *causal* explanations. Despite the notoriously slippery character of the word "causal," this restriction disallows certain proposals from the outset. For one thing, as we have already seen, not all "why" questions are questions about the cause of something or other. Thus, if one asks, "Why did the rat run to the food box

[3]Marx and Freud suggest that the reasons we offer in support of our beliefs are invariably no more than "rationalizations" of them. Their true causes, unbeknownst to most of us, are economic (Marx) or psychosexual (Freud).

[4]For one thing, it construes numbers as having classes as members. But for many people, numbers are not "made up" of anything, let alone classes.

[5]In fact, several different versions satisfy the Peano Axioms. One might rest content with this result, or go on to formulate additional requirements in order to eliminate some of these versions.

in the maze?" the answer might be, "In order to get food" or "Because he was hungry." The first answer can hardly be regarded as giving the cause of the animal's running to the food box, because getting to the food occurs *after* the running. Whatever the analysis of the word "cause," no cause occurs later than its effect. Therefore, the answer, "In order to get food," is not a causal explanation of the rat's running behavior. On the other hand, the answer, "Because he was hungry," is a causal explanation of his behavior. For presumably the rat's hunger preceded his running. Of course, fixing on the causal explanation as the object of our analysis does not preclude the possibility that the other type indicated (sometimes called a *motive* explanation) can eventually be assimilated to it. We shall in fact have something to say about the assimilation of this and other (allegedly "noncausal") types of explanation later in the chapter.[6] For another thing, restricting our attention to causal explanations limits our discussion (at least initially) to the explanation of given *events*.

Now what requirements should we place on the adequacy of an analysis of the concept of causal explanation? What intuitions about explanation should such an analysis embody? Three intuitions suggest themselves as particularly central. The first is that our analysis conform to the patterns of explanation regarded as paradigms in the various established sciences. If, for example, a given analysis resulted in our refusal to call the account of bodies in motion offered by classical (Newtonian) mechanics "explanatory," it would have to be rejected. The second is that whatever the analysis, an explanation must have empirical content. The notion of "empirical content" is difficult to make precise, as we shall see in a moment. What is intended is simply that any explanation may be put (possibly in indirect ways) to empirical test. Appeals to God's will, for instance, although satisfying to many people, are not generally held to be explanatory; that the Lisbon earthquake occurred because God willed it is not really an assertion open to scientific investigation. The third, and in many ways most important, requirement is that an explanation indicate why, given certain antecedent conditions, the event to be explained *could* have been expected to occur. In other words, an explanation shows how, given the facts of the case, the event in question had to occur; it is in some sense (to be made clearer) necessary relative to its antecedents. To put it still another way, that the rat was hungry provides an explanation of his running

[6]Certain types, for example, mathematical explanation, will be excluded from the discussion without further comment.

when, taken together with other relevant data, we could have predicted the running on the basis of the hunger.

2. The "Humean" Account

The rudiments of an analysis that satisfies our requirements admirably can be found in David Hume's *A Treatise of Human Nature* (1739). Hume's concern was not primarily with the concept of explanation, but rather with the nature of the causal connection between events. He argued that this connection was not *necessary*, in the sense that we can consistently suppose a given event to be followed by any one of a number of different events. Equivalently, to say that one event causes another is not to say that a statement describing the second follows logically from a statement describing the first. Hume illustrates the point in another of his works, the *Enquiry Concerning Human Understanding:*

> When I see, for instance, a billiard-ball moving in a straight line toward another; even suppose motion in the second ball should be suggested to me, as a result of their contact or impulse; may I not conceive, that a hundred different events might as well follow from that cause? May not both these balls remain at absolute rest? May not the first ball return in a straight line, or leap off from the second in any line or direction? All these suppositions are consistent and conceivable.

Instead, this connection is to be understood in terms of the regular succession of certain sorts of events. To say that an event of kind *A* caused an event of kind *B* is to say that whenever an event of kind *A* occurs, it is invariably followed (as a matter of fact, not of logic) by an event of kind *B*. Thus, to return to the billiard balls, we say that the impact of one caused another to move off at a certain velocity because these two sorts of events are regularly conjoined.

Most of Hume's examples are drawn from the seventeenth century science of mechanics. But the central feature of his account is already prominent (though implicit) in the concept of cause which Herodotus employs in his *History*. At one point, for instance, various "causes" of the Nile's rising every summer are suggested, in particular that the phenomenon is produced by the seasonal blowing of the Etesian winds. But, Herodotus goes on to object,

> . . . if the Etesian winds produced the effect, the other rivers which flow in a direction opposite to those winds ought to present the same phenomenon as the Nile, and the more so as they are all smaller streams, and have a weaker current. But those rivers, of which there are many both in Syria and Libya, are entirely unlike the Nile in this respect.

To say that the Etesian winds cause the Nile to rise is to say tha events of the two sorts are regularly conjoined. Since in this case nc such regularity obtains—Herodotus alludes to a number of counter instances—we can conclude that the Etesian winds do not in fac cause the Nile to rise. The cause of this phenomenon must be sough elsewhere.

Hume's account of "cause" adapts quite easily to "explana tion." Given the preliminary restriction of our discussion to causa explanation, we can say that to explain the occurrence of som event is to say what caused it. Thus, on Hume's account, to explai the occurrence of some event is to exhibit it, together with certai antecedent events, as an instance of a regularity or, to use a provi sionally equivalent expression, general law. We are not home free "Instance" and "general law" still require analysis, and they wil prove to be extremely troublesome. But we have made progress. Ir particular, it should be clear how the rudimentary analysis giver meets our three original requirements. First, it appears to conforn to the patterns of explanation regarded as paradigms in the variou established sciences. We have already mentioned Herodotus. Per haps more to the point is the kind of explanation typically set ou by Newton. The motion of a particular planet is explained b referring to certain regularities or laws that obtain in the motion o all planets, indeed of all objects whatsoever. Second, an explanatio on the "Humean" account has empirical content. To put an expla nation to empirical test is to test the laws or regularities on whic it rests. Finally, given these laws and certain additional informatio the occurrence of the event in question could have been predicte (for the laws assert that given such and such antecedent condition it invariably follows); in this sense, it was to be expected.

This account can be elaborated in much greater detail, esp cially with an eye toward making its logical structure explicit. Th classic contemporary presentation is in a paper by Carl Hempe and Paul Oppenheim, "Studies in the Logic of Explanation." We will approach it by way of a simple (and somewhat artificia example that might be taken as a rough approximation to mor complex scientific explanations.[8]

Someone asks "Why did thread *T* break?" A plausible answer i

[7] *Philosophy of Science,* 15 (1948), pp. 135–75.

[8] The example is Karl Popper's, *The Logic of Scientific Discovery* (Londo Hutchinson & Co. (Publishers) Limited, 1959), pp. 59ff. Many points in th present discussion are due as much to Popper as they are to Hempel and Oppe heim.

that a heavy weight was suspended from it. Now on the "Humean" account, if the first of these events was caused by the second there is an appropriate generalization or law "covering" them,[9] perhaps in the following way: whenever a heavy weight is suspended from a thread it will break. Given this (presumed) law and the fact that a heavy weight is suspended from thread T, it *follows* that thread T breaks. In terms of the law and the fact, the thread's breaking has been explained.

The structure of this explanation can be made still more explicit. It has five parts.

1. For every thread of a given structure S (that is, of a certain material, thickness, and so on), there is a characteristic weight W, such that the thread will break if a weight exceeding W is suspended from it.
2. For every thread of the kind S, the characteristic weight $W = K$.
3. T is a piece of thread of the kind S.
4. B, which has a weight greater than K, is suspended from T.
5. Therefore, T breaks.

This explanation has the form of a deductive argument in which (5) is a valid consequence of premises (1) through (4). In a more technical vocabulary, (5) is called the *explanandum*, a statement describing the event to be explained. (1)–(4) are called the *explanans*, or explaining statements. (1) and (2) are general laws, (3) and (4) are statements of the antecedent conditions obtaining. Presumably all these statements have empirical content and, further, are true. To explain the occurrence of an event, then, is to derive a statement describing it from other true and empirical statements, at least one of which is a general law. To put it more briefly, explanation is deductive subsumption under general laws.

It must not be imagined that all the explanations one sees, or hears, look like the one just set out. Most explanations encountered are analogous to what logicians call *enthymemes*. An enthymeme is an argument with suppressed premises. For example, the argument that Nixon is a citizen because he is president is an enthymeme. The statement "Nixon is a citizen" does not follow from the statement "Nixon is president" alone. To see this, consider a counterexample: "Nixon is president; therefore, he is a Democrat." This has the same form as the initial argument, but is invalid. Rather than conclude that the initial argument is invalid, one would likely assume that what the speaker has in mind is the following: "Nixon

[9]Hence this account is also known as the "covering law model" of explanation.

is president, and if he is president then he is a citizen; therefore, he is a citizen." This argument is valid and includes the formerly suppressed premise, "If Nixon is president, then he is a citizen." Likewise, explanations are usually abbreviated editions of valid arguments.[10] What is usually left tacit in an explanation is the reference to the law or generalization. Such was the case in the passage cited from Herodotus. The point is that anything which is offered as an explanation must be capable of being expanded into the above form.[11]

One more point can be made briefly. It is that this "law-deductive" account of explanation meets our basic requirements. This should come as no great surprise; our analysis was framed so as to incorporate directly the intuitions that these requirements involve. Further, we have already seen how the rudimentary "Humean" account from which it springs satisfies the requirements. But perhaps it is worthwhile to review them once again. First, it seems to conform to established scientific practices. This is not to say that every accepted scientific explanation takes the form of a deductive argument containing at least one general law and the conclusion of which is a statement describing the event to be explained. It is to say that all such explanations can be put into this form (that is, this form makes explicit the underlying structure that they all share). Second, it provides for the requirement that an explanation can be put to empirical test; trivially, the *explanandum* itself constitutes such a test. Finally, the *explanans* offers logically conclusive grounds for expecting the event to be explained and thus satisfies the third requirement.

3. Inductive-Statistical Explanation

The account of explanation we have just sketched has two central features: the *explanandum* is a logical consequence of the *explanans* and the *explanans* includes at least one general law. These two features are related in a way that we will now make clear.

So far we have considered explanations resting on laws which are of universal form. In all cases in which such and such conditions

[10]Hempel calls such abbreviated editions "explanation sketches."

[11]Although perhaps not by the person who advances the "explanation sketch." The string broke because a heavy weight was suspended from it, but not many of us could state the sophisticated laws that in fact (and in contrast to our very simplified example) relate the two events. Jones dies because he took arsenic; is the explanation any the less satisfactory because the person who advances it cannot state the requisite pharmocological laws?

are realized, so and so kind of event will result—"For every thread of a given structure *S* there is a characteristic weight *W*, such that the thread will break if a weight exceeding *W* is suspended from it" —or, all objects of a certain kind have a particular property—"All mammals suckle their young." Many scientific laws (for example those of classical mechanics) have this character. But there are also laws which have a different, statistical, character. They assert that when such and such conditions are realized, there is a certain statistical probability that so and so kind of event will result, or that a certain percentage of a given population has a particular property. For instance, it is a fundamental law of genetics that in a cross of heterozygotes (animals with one dominant and one recessive gene) one-fourth of the next generation will be homozygote (unmixed) dominant, one-half will be heterozygote, and one-fourth will be homozygote recessive.

This contrast between laws of universal form and statistical laws is important for the following reason. In explanations resting on statistical laws, the *explanandum* is *not* a logical consequence of the *explanans* as is the case when the laws are of universal form.

Suppose we wished to explain Jones' having had a heart attack on the basis of the statistical generalization (taken as true) that 90 percent of the persons in his age group have heart attacks.[12] The premises of our putative explanation would then be:

(1) 90 percent of the persons in age group *G* have heart attacks.

(2) Jones is in age group *G*.

But it should be clear at once that the conclusion we want,

(3) Jones has a heart attack,

is not a logical consequence of (does not follow from and is not deductively implied by) these premises, which is to say that the two premises might be true and yet "Jones has a heart attack," be false.

It might be thought that although (3) does not follow from premises (1) and (2), another conclusion,

(3*) the probability of Jones having a heart attack is 90 percent,

does follow. But this is not case: (3*) does not follow any more than does (3) from (1) and (2). To see this consider a parallel argument, both of whose premises we assume to be true.

(1′) 30 percent of the persons of nationality *N* have heart attacks.

[12]The example is Israel Scheffler's, *The Anatomy of Inquiry* (New York: Alfred A. Knopf, Inc., 1963), p. 35.

(2′) Jones is of nationality *N*.

Therefore,

(3*′) the probability of Jones having a heart attack is 30 percent.

Quite obviously the probability of Jones having a heart attack cannot simultaneously be 90 percent and 30 percent. Although both sets of premises are presumed true, their respective conclusions are incompatible, and this would not be possible if these conclusions followed from the premises.

In cases of this kind, where the laws are of a statistical character, the *explanans* does not deductively imply the *explanandum*, even if the *explanandum* itself is a probability statement like (3*) or (3*′). Rather, it confers a certain "likelihood" on it. Given the statistical generalization that 90 percent of the persons in age group *G* have heart attacks and the fact that Jones is in age group *G*, then it is highly likely that Jones will have a heart attack. Jones' having a heart attack is not likely per se, but relative to this fact and this generalization. At the same time, given the statistical generalization that 30 percent of the persons of nationality *N* have heart attacks and the fact that Jones is of nationality *N*, then it is less likely that Jones will have a heart attack. As relativized to their premises in this way, the two conclusions are no longer incompatible.

We must be very careful to distinguish between the concepts of *likelihood* and *statistical generalization*. This is all the more difficult because the term "probability" is habitually used in speaking of each. A statistical generalization is a statement to the effect that a certain percentage of cases having feature *F* will have feature *G*: 90 percent of the persons in a particular age group have heart attacks. We might also say that a statistical generalization states the relative frequency with which certain kinds of events have certain kinds of outcomes. Likelihood, on the other hand, is a relation between statements. It is a measure of the degree of support which some statements confer on others, in particular the degree of support which the *explanans* confers on the *explanandum*. Likelihood is sometimes called "*inductive* probability" to distinguish it from "*statistical* probability," and arguments in which the conclusion is not a logical consequence of the premises but is supported by them to a greater or lesser extent are called "inductive arguments."

There are two more points which must be noted, for they are sources of frequent misunderstandings. The first point is that the mere presence of statistical generalizations in an argument does not render it inductive, nor does the absence of statistical generalizations from an argument render it deductive. What is the case is that arguments *whose conclusion describes an individual event* and which contain statistical generalizations are inductive. Otherwise, there are deductive arguments that contain statistical generalizations, although they explain not individual events but what we might call "mass" events.[13] Given the (presumed) generalization that 6 percent of all American cigarette smokers get lung cancer and the fact that there are 100,000,000 American cigarette smokers, then we can *deduce* that 6,000,000 American cigarette smokers will get lung cancer.

On the other hand, there are arguments that one might want to take as "inductive" (perhaps in an attenuated sense), although they contain no statistical generalizations.[14] For example: let "dx" stand for "x has a certain disease," "px" stand for "x has a high pulse rate," "bx" stand for "x has high blood pressure," "rx" stand for "x has severe pains in his respiratory tract," and "fx" stand for "x has a fever." Further, let us assume that the following four statements are true:

1. For every x, if dx then px.
2. For every x, if dx then fx.
3. For every x, if dx then bx.
4. For every x, if dx then rx.

In addition, suppose that a given person, Jones, is such that he has f, p, b and r. Then all of this information in the hands of a physician might be regarded as symptomatic evidence that Jones has a certain disease d. This inference is not deductive and, as stipulated, contains no statements of a statistical character.

The second point to be noted is that we have to distinguish between the *form* of the law in question, that is, whether it is universal or statistical and the degree to which it is supported by the available evidence.

The fact that all laws are more or less probable relative to the evidence we have for them is independent of the kind of claim particular laws make, for example, whether they ascribe a property to all members of a reference class ("All crows are black") or just to a specified proportion of it ("90 percent of all swans are white").

[13]The example that follows is May Brodbeck's, "Explanation, Prediction, and 'Imperfect' Knowledge," in *Minnesota Studies in the Philosophy of Science,* Vol. III, eds. H. Feigl and G. Maxwell (Minneapolis: University of Minnesota Press, 1962).

[14]Scheffler, *The Anatomy of Inquiry,* p. 40.

Thus the fact that a law is merely probable, in the sense that our evidence for it is not logically conclusive, does not make explanatory arguments in which it figures inductive. What matters is not the evidence we have for a law but, once again, the kind of claim the law makes; the crow and swan generalizations make different kinds of claims, but presumably the evidence for the one is not any better than it is for the other.

The original Hempel and Oppenheim paper, to which we have already referred, emphasizes the deductive character of explanatory arguments. Yet if we admit statistical generalizations of the kind mentioned in our example, and agree that our aim is the explanation of individual events, then the deductive character of explanatory arguments cannot be preserved. Are we forced to say that inductive accounts do not qualify as genuine explanations?

Many philosophers of science answer "yes." For them, to be told that Jones had a heart attack because 90 percent of the persons in Jones' age group have heart attacks is a sorry explanation—indeed, none at all. Thus Israel Scheffler writes:

> [To be told this] is, normally, to leave out of account just what we wanted to know, that is, why Jones had an attack, how he was differentially constituted and how he differentially acted. To say that there is a high incidence of [heart] attacks in [Jones'] age, occupational, and geographical group is compatible with saying we do not *understand why* [Jones] was afflicted with a [heart attack].[15]

To put the matter briefly, purported explanations, in which the event to be explained is only inductively connected with the purported *explanans,* provide at most a *ground* for our belief that Jones had a heart attack, but do not *explain* the attack.

On the other hand, there are philosophers who do want to say that inductive accounts qualify as genuine explanations. They take their stand largely on the fact that statistical generalizations are coming to play an increasingly important role in the development of science—for example, in statistical mechanics, quantum theory, and genetics, and perhaps most prominently in the social sciences. To mention but one instance, Durkheim's celebrated study of suicide proceeds to an explanation of cases by way of statistical generalizations about certain social types and classes. Further, the accounts in which these generalizations figure do seem to be explanatory, although perhaps not so securely or so satisfactorily as strictly deductive arguments.

[15]*Ibid.*, p. 35.

The nub of the controversy is one of the requirements we originally placed on an analysis of the concept of explanation; namely, that an explanation indicates how, given certain antecedent conditions, the event in question could have been expected to occur. Whether one wishes to maintain intact or relax the deductive requirement depends on how one construes this notion of "expectation."

The brief for the deductivists is very well stated by Alan Donagan:

> If your task is to explain why a given event *E* occurred, rather than did not occur, then your *explanans* must exclude the possibility that *E* did not occur; but if your explanation is not deductive, i.e., if its *explanans* does not logically entail its *explanandum*, then it will not exclude that possibility and so will not explain why *E* did not occur.[16]

To show that the event *E* was to be "expected" is to exclude the possibility that *E* did not occur. It is to show why, in some sense, *E* was necessary or had to happen, relative to certain antecedent circumstances.

The "inductivists" (these who count inductive as well as deductive arguments explanatory), on the other hand, interpret "expectation" less stringently, as requiring only that the *explanans* must exclude to a high degree of probability the possibility of *E*'s not occurring. Once again, they stake their case principally on what they take to be the actual situation in science. The first of our requirements on the adequacy of any analysis of the concept of explanation, it will be remembered, was that it conform to actual scientific procedures. Insisting on the importance of this requirement, the "inductivists" adapt "expectation" accordingly.[17]

How is the issue to be resolved? In addition to those already considered, several different factors are relevant.

[16]"The Popper-Hempel Theory Reconsidered," in *Philosophical Analysis and History*, ed. W. Dray (New York: Harper & Row, Publishers, 1966), p. 132; Cf. Brodbeck, *Minnesota Studies*, p. 239: "Either the explanation is deductive or else it does not justify what it is said to explain."

[17]Interestingly, the "inductivist"/"deductivist" controversy has at times figured as an internal question in various branches of science. For example, some of the debate concerning the adequacy of quantum theory is between those who contend that it does not meet a satisfactory standard of "expectation" and those who maintain that in view of the successes of the quantum theory that standard must be revised.

Hempel himself has found reasons to press for an "inductivist" position. Cf. his paper "Deductive-Nomological vs. Statistical Explanation," in *Minnesota Studies in the Philosophy of Science*, Vol. III.

One factor that can be no more than suggested here has to do with the development of inductive logic. There is, to date, no adequate and widely accepted characterization of the inductive probability that a given *explanans* confers on its *explanandum*. The situation contrasts rather sharply with that in deductive logic, where the corresponding notion of logical consequence is well defined. The prospects for the "inductivist" position hinge in large part on the eventual success or failure of inductive logic.[18] We will have more to say about this in the chapter on confirmation, in which some of the problems confronting the development of such a logic are raised.

A second factor, again already alluded to, concerns the extent to which science aims to explain and predict individual events. For as we have seen, the use of statistical generalizations inevitably forces an inductive argument only in cases in which we are concerned to explain individual events. In fact, some philosophers[19] urge that science (that is, *pure* as opposed to *applied* science) is always concerned with general facts, not particular events.

A third factor is simply this: perhaps neither the "deductivist" nor the "inductivist" positions as they stand provide an adequate account of explanation. Let our question be as before—"Why did Jones have a heart attack?" Someone might answer that it matters little whether 90 or 100 percent of the persons in Jones' age group have heart attacks. Neither generalization serves to explain the event in question. Roughly, asking for an explanation of a heart attack is asking for the physiological conditions, and so on, which *cause* or *produce* a heart attack. But there is no reference to these in the statement that some or even all the persons in a specified age group have heart attacks.

This challenge has two different sorts of response. One is to dismiss it as a metaphysical relic, disposed of long ago by Hume. When we ask for the "cause" of something, we are asking for *no more than* a generalization under which it may be subsumed. There are no "causes" over and above regularities that obtain in nature; events have no additional "inner springs."

[18]The work of Rudolf Carnap in this respect is of special importance. See his article, "The Aim of Inductive Logic," in *Logic, Methodology and Philosophy of Science*, eds. E. Nagel, P. Suppes, and A. Tarski (Stanford: Stanford University Press, 1962), pp. 303–18.

[19]From Aristotle who argued that scientific knowledge is always of universals to Bertrand Russell: ". . . Science, though it starts from observation of the particular, is not concerned essentially with the particular, but with the general. A fact, in science, is not a mere fact but an instance." *The Scientific Outlook* (New York: W. W. Norton & Company, Inc., 1931), pp. 57–58.

A second response is to amend the present account of explanation. To explain some event is not simply to subsume it, deductively or inductively, under some true generalization, but to subsume it under a "causal law." A distinction between causal laws and other sorts of generalizations suffices, it might be added, to make clear how "spurious" diverge from "genuine" explanations or, more guardedly, to distinguish between "causal" and "noncausal" explanations. For example, "all persons in age group *G* have heart attacks" is not a causal law. In section four of this chapter, various issues surrounding the concept of a law are taken up, and we will discuss this distinction in greater detail.

A number of promissory notes have been given lately. We must now set to work making them good.

4. Laws and Conditional Statements

We can begin with the concept of a law. The first thing to notice is that "law," like "cause" and "explanation," is used in a variety of ways, many of them only vaguely related.[20] In fact, we ourselves have used "law" (or "general law") interchangeably with "generalization," "regularity," and "statement of universal form." Before proceeding much further, these various expressions will have to be sorted out.

Following what is perhaps the most widespread custom, let us say that laws are *statements* that express *regularities*. The properties of laws in which we will be interested will be properties of statements. It is important to keep this in mind, for confusion results when one fails to distinguish between laws as statements and laws as what certain statements express. "Generalization" is ambiguous in exactly the same way: for reasons of clarity, we construe generalizations also as statements and not as what statements of a certain kind express. The question, then, is to decide whether there is a distinction to be made between laws and generalizations, as statements of different types.

One further preliminary. We are going to restrict our attention to the simplest case, where a law has the form of a nonstatistical *generalized conditional* statement. In other words: where a law has the form "For any (every, all) object(s) x, if x is such and such then x is so and so." Such nonstatistical generalized conditionals may also be said to be statements of *universal form*. For they assert that every object (or event or condition) that has certain properties also has certain other properties. One of the laws we have mentioned previ-

[20]In the case of laws of nature and laws of the land, this relation seems to be very vague indeed.

ously is a case in point: "For every piece of thread T, there is a threshold weight W, such that if an object is suspended from T and exceeds the threshold W, then the thread will break." It should be noticed that mathematically phrased laws merely appear to have a different form. Consider the mathematical relationship about specific gravity: "$G_s = W/V$." What is suppressed in the original mathematical statement is the reference to all objects, that is, its generality.

Let us assume that laws are statements of universal form.[21] At the same time, it does not seem to be the case that all statements of universal form are laws (for example, "All animals are men"). This claim can be defended by considering additional criteria that statements of universal form must satisfy to be counted as laws.

One criterion is that laws are universal statements that have empirical content. A use to which this criterion is frequently put is to rule out nonsense laws and spurious explanations based on them. For example, one might want to say that the universal statement "All glubbifiers are irascible" is patent nonsense, hence no law, and any "explanation" in which it figured would be unacceptable. It is not at all clear, however, that the line between "sensical" and non-sensical statements coincides with the (for the moment presumed) line between statements that have empirical content and those that do not. It is even to be wondered whether there is *any* effective way of separating sense from nonsense. For many of his sixteenth century contemporaries, Copernicus' heliocentric hypothesis was the "Copernican Paradox." For one thing, it was virtually part of the meaning of the word "earth" that this planet be the center of the universe.

The criterion of empirical content is also intended to rule out, for example, "All bachelors are unmarried men," as a law, and the following as a genuine explanation of the fact that Jones is an unmarried man: "Jones is a bachelor and all bachelors are unmarried men; therefore, Jones is an unmarried man."

Indeed, we would undoubtedly be very hesitant to accept this as an *explanation* of the fact that Jones is an unmarried man. And it is tempting, as a result, to distinguish between laws and non-laws in terms of the empirical content of the former. To use an expression introduced in the previous chapter, *analytic* statements (that is, statements true by virtue of the meanings of the words they contain), even of universal form, are not *laws*. There is something

[21]We said earlier that certain laws are of statistical or probabilistic form. Probabilistic laws raise their own problems, but none of these affect the points to be made in this section.

circular about the "explanatory" arguments into which they enter. Thus, to consider another example:

1. All paranoids suffer delusions of grandeur.
2. Jones is a paranoid.

Therefore,

3. Jones suffers from delusions of grandeur.

If we assume that the definition of the word "paranoid" includes as part of its meaning the expression "suffers from delusions of grandeur" (that is, (1) is analytic), this "explanation of Jones' suffering delusions of grandeur because he is paranoid does not appear to advance our understanding.

Despite the intuitive appeal of these remarks, however, we must be very careful. The notion of "empirical content" and the attendant analytic/synthetic distinction are unclear. Admittedly in the cases just considered, the intended analytic/synthetic, empirical/nonempirical contrasts are fairly easy to draw.[22] But cases of this kind are by and large trivial. In more complicated cases of the sort one typically encounters in science, a great many principles are involved in the explanation of particular events. Perhaps one can distinguish between these as having "more" or "less" empirical content; within the context of some specified theory, one can make the distinction sharper still. If the line of argument sketched in the final section of the previous chapter was sound, however, a general distinction between statements that have empirical content and those that do not, for example, between the laws of mathematics and the laws of physics, is at least problematic (once again excepting trivialities like "All bachelors are unmarried men," which are of negligible scientific and systematic importance). We do not have to abandon our earlier requirement that explanations have empirical content. But we should be alert to the possibility that it might need modification.

A second criterion laws might be expected to fulfill is that they be true. Thus, "All swans are white" does not count as a law (assuming provisionally that it satisfies the other criteria) because there are black swans; we cannot explain a particular swan's being white on the basis of it.

[22]There is a complication in the "paranoid" case. Very plausibly, we could say that suffering delusions of grandeur was a *symptom* of or criterion for paranoia and that although in virtue of this fact there was a kind of "meaning" connection between them this did not preclude "All paranoids suffer delusions of grandeur" from functioning as a law.

This criterion seems even more intuitive than the first. Surely we would not countenance false laws of nature. But there are difficulties with it as well.

The first is that many statements commonly regarded as "laws," for example, those of classical mechanics, hold only approximately. They are not, in all strictness, true. If we insist that explanations essentially involve laws and that laws must be true, we are faced with the unpleasant choice of saying either that, for example, Galileo did not *really* explain certain facts about the behavior of objects on inclined planes, or that when he made them his statements were true but now no longer are. Perhaps a less painful move is to relax the requirement that laws be true. There are also reasons (to be discussed more fully in the chapter on confirmation) for maintaining that at least some laws of nature are neither true nor false. Rather, they serve as general principles, somewhat on the order of moral rules,[23] whose function is to provide a theoretical framework with respect to which research can be carried out and empirical generalizations tested. To put it in a slightly different way, at least some laws of nature seem to have a primarily normative, not descriptive, function.

In any case, there are reasons for leaving it open for now whether laws must be true or not. We can do this by restricting our discussion to *lawlike* statements. Statements are lawlike if they satisfy all the other criteria for laws, independent of whether or not they have a truthvalue.

The third, and most important, of these criteria can best be taken up by way of example.[24] Consider the two statements:

1. All persons sitting on a certain bench in Boston are Irish.
2. Any body subject to no external forces maintains a constant velocity.

There are many (interrelated) differences between these two statements that might lead us to call the second, but not the first, "lawlike," although both are of universal form. Recall that it is not in virtue of their form that laws are to be distinguished from non-laws.

To begin with, the first statement seems to be an *accidental* generalization, about what in fact happens to be the case, whereas the second is a law of nature. It does not follow from the first that

[23] We do not mean to suggest that they are like moral rules in all respects.

[24] A classic contemporary discussion of many of the points that follow can be found in Nelson Goodman's book, *Fact, Fiction, and Forecast* (Indianapolis, Indiana: The Bobbs-Merrill Company, Inc., 1965).

if someone *were* to sit on a certain bench in Boston he *would* be Irish, nor could we very well explain someone's being Irish by referring to the fact that he was sitting on that bench. But it does follow from the second that if a body *were* subject to no external forces it *would* maintain a constant velocity, and it does (intuitively) explain a body's having a constant velocity to be told that it is subject to no external forces.

Statements of the form "If *A* were (had been) the case, then *B* would be (would have been) the case," where in fact *A* is not (has not been) the case, are called *counterfactual conditionals.* The present suggestion is that lawlike statements support counterfactual conditionals, while nonlawlike statements do not. In turn, the ability to support counterfactuals (as they are called for short) is linked to the predictive, and hence explanatory, force that laws in contrast to non-laws have.

Conditional statements quite generally allow us to speak about potential (as distinct from actual) events, cases in which should *A* occur, *B* would also. There is thus a sense in which lawlike statements, insofar as they support conditionals, possess a generality that nonlawlike statements do not. The accidental generalization, "All persons sitting on a certain bench in Boston are Irish," is equivalent to the finite conjunction of statements, "Pat is sitting on a certain bench in Boston and is Irish," and "Fitz is sitting on a certain bench in Boston and is Irish," and so on, whereas the lawlike statement, "Any body subject to no external forces maintains a constant velocity," is unrestrictedly about all objects whatsoever. This is perhaps another way to indicate its explanatory force.

A distinction between laws and generalizations on this basis has recently been given a more than academic prominence in the controversy surrounding cigarette smoking and health. Spokesmen for the tobacco industry are inclined to say that statements expressing a correlation between cigarette smoking and, for example, lung cancer are merely accidental generalizations. A high statistical correlation should not be mistaken for a causal connection; generalizations linking cigarette smoking and lung cancer do not support conditionals to the effect that if someone were, for example, to smoke two packs of cigarettes daily for a specified period of time, the chances of his getting lung cancer would be such and such. The Public Health Service, on the other hand, tends to grant the status of laws to these same generalizations. Which is to say, at least, that they do support the appropriate conditionals. Waiving difficulties with the notion of "confirmation" for the moment, the P.H.S. might want to stake its case further on the extent to which projections made on the basis of the purported laws were in fact realized. Of course, this still would not force the tobacco industry to say that a "causal" connection had been demonstrated; other criteria of lawlikeness, they could insist, remain to be satisfied.[25]

We take it, then, that it is apparently a necessary (if not also a sufficient) criterion for being a lawlike statement that a statement of universal form support conditionals, in particular counterfactual conditionals. The usefulness of this criterion depends in part on the sureness of our grip on the notions of "support" and "counterfactual conditional."

One way of analyzing such conditionals, in terms of the so-called "material conditional" of classical logic, will not work. The material conditional—"If *P* then *Q*"—is so defined as to be true whenever "*P*" is false or "*Q*" is true. Since the antecedents of counterfactual conditionals are always unrealized, that is, "*P*" is always false, counterfactuals construed as material conditionals would always be true. But this is not acceptable, as can be seen by comparing the two counterfactuals.

1. If any body were subject to no external force, it would maintain a constant velocity.
2. If any body were subject to no external force, it would *not* maintain a constant velocity.

On the model of the material conditional, both counterfactuals are true, although their consequents are incompatible! A related way of rejecting the material conditional analysis is to point up the importance of distinguishing between conditionals that are vacuously true (true simply because their antecedents are unrealized) and those that are not. The material conditional analysis obviously does not allow us to do this.

More generally, many conditionals simply are not truth-functional. We may know the truthvalues of their component statements

[25]The cigarette smoking lung cancer controversy illustrates in a very sharp way almost all the issues raised in this chapter. The Public Health Service, for example, is explicitly "inductivist" about explanation; it maintains that a high statisical correlation can, when subjected to a variety of tests, be used to assert a causal connection, that is, can be used to explain individual instances of lung cancer, and so on.

Both sides of the controversy are also quite explicitly aware that in many respects the issue is philosophical: How is the concept of cause to be analysed? See the very interesting discussion in the Report of the Advisory Committee to the Surgeon General of the Public Health Service, *Smoking and Health* (Princeton, N. J.: Van Nostrand Reinhold Co., 1964), pp. 19–21 and *passim*.

and, from this information alone, be unable to determine the truth-value of the compound. For example, I may know that "Willie Mays played in the American League" is false, as is "Willie Mays hit four hundred," without knowing whether or not Mays would have hit four hundred had he played in the American League.[26]

A more promising suggestion is that a conditional is like an enthymematic argument, one premise of which is the antecedent of the conditional, the conclusion of which is the consequent. What is omitted, and what serves to connect antecedent with consequent, is simply the law (or laws) that we have taken to support it. A (vastly oversimplified) case in point: "Had the match been struck it would have lighted" is a counterfactual, the consequent of which is derivable from the antecedent with the help of certain laws about matches being struck and lighting.

The proponents of this suggestion maintain that, intuitively, there is a "connection" between maintaining a constant velocity and being acted on by no external forces, whereas there is no such "connection" between sitting on a certain bench in Boston and being Irish. It is this connection that allows for valid predictions, and hence, explanations. But the connection between antecedent and consequent in a true conditional is none other than that which a law provides. From the antecedent and the law, we can derive the consequent. It is in this way that laws serve to support conditionals. Similarly, the *necessity* of a match's lighting when struck is simply a function of the fact that it follows from the law as a logical consequence.

There is a circle here, of course. We have distinguished between laws and generalizations by claiming that only the former supported counterfactual conditionals, and we have analyzed such conditionals in terms of the lawlike connection that exists between their antecedent and consequent. The circle does not undermine the distinction suggested. But it does indicate that we do not understand the distinction very well unless the circle can be broken. One way in which to break the circle is to say that lawlike statements, in contrast to accidental generalizations, are *confirmed by their instances*. A discussion of this proposal must await the chapter on confirmation.[27]

[26]The example is Robert Stalnaker's; his analysis of conditional statements will be mentioned shortly.

[27]Another, extremely interesting, analysis of conditionals has recently been advanced by Robert Stalnaker. See his paper, "A Theory of Conditionals," in *Studies in Logical Theory*, ed. Nicholas Rescher (American Philosophical Quar-

One could also argue, as have some philosophers, that additional criteria of lawlikeness are required if laws are to play the explanatory role assigned to them. Support of counterfactual conditionals is not enough. For example, even if it were a conditional-supporting generalization that every time the sky is red in the morning it rains in the afternoon,[28] we would not explain the fact that it is raining in the afternoon by pointing to this purportedly "lawlike" statement and the fact that the sky was red this morning. Other philosophers reject this sort of distinction between explanatory laws and nonexplanatory generalizations, contending that it is rooted in naive (unexamined) causal and teleological intuitions. For them, the only distinction to be made is that between laws and accidental generalizations, by means of the counterfactual conditional criterion.

But consider these two statements:

1. All pieces of copper wire at $-270°$ C are good conductors.
2. All unicorns are fleet of foot.

Both of these statements are unrestricted universals, and presumably both would support (insofar as they implied) the corresponding counterfactual conditionals. Take it as (nonanalytically) true that if anything were a unicorn it would be fleet of foot. Still, it is unlikely that anyone would regard the statement about unicorns as a law. Nor can the two statements be distinguished by virtue of the fact that the first but not the second has positive instances. We can easily suppose that no copper has been checked out at $-270°$ C. The difference, rather, seems to lie in the fact that the copper-conducting statement is systematically connected with a larger theoretical framework, whereas the unicorn-fleeting statement is not. To put it in a slightly different way, what enables us to call the first but not the second statement "law like" is the *position* of the

terly Monograph Series, 1968). Stalnaker is concerned with the analysis of conditional statements generally, not primarily, as we have been, with counterfactuals. He rejects both of the analyses suggested above, and proposes instead an analysis in terms of possible or hypothetical states of affairs.

Stalnaker's analysis has a great deal to recommend it. It does not, as he himself notes, however, allow us to bypass the circle described in analyzing counterfactual conditionals. We may know the truth-conditions for conditional statements (that is, the conditions with respect to which they are evaluated) without knowing whether particular conditionals are true or false. And presumably those counterfactuals, at any rate, are true which are supported by laws in the way we suggested originally.

[28]That is, even if this were a generalization that held without exception and had unrestricted application.

first in the existing body of knowledge. In particular, whether or not it has any positive instances, it will be granted law status if it *follows from* other laws and theories. That copper at —270° C is a good conductor, for instance, ostensibly is a consequence of the law that all copper is a good conductor.

On the other hand, even well-confirmed unrestricted universals will not count as laws if they are not systematically interconnected with a larger theoretical framework. What leads us to reject the red sky in the morning/rain in the afternoon generalization as a law, or an *explanatory* law, is that it is an isolated assertion having no apparent theoretical ramifications.[29]

This point can be elaborated. It is not enough that a particular universal statement be exceptionless to qualify as a law. It must also be the case that the theoretical framework to which it is connected does not permit exceptions to it. Certainly nothing in current scientific theory precludes, for example, that anyone sitting on a certain bench in Boston is not Irish, nor that unicorns are not fleet of foot. Similarly, whether there are laws linking cigarette smoking to lung cancer depends on how the statistical generalizations correlating them relate to a larger body of scientific theory.[30] Generalizations that are not systematically connected with other laws are merely accidental, they rule out potential instances (for instance, a non-Irishman sitting on that bench in Boston) that scientific theory otherwise permits.

5. Explanation and Prediction

One of the criteria of adequacy for a satisfactory account of explanation is that an explanation indicates why, given certain antecedent conditions, the event to be explained could have been *expected* to occur. This amounts to requiring of any explanation that it be a potential prediction; in this precise sense, the event to be explained could have been expected to occur.

The account of explanation under consideration seems to satisfy this criterion admirably. To explain the occurrence of some event is to derive a statement describing it from other statements, at least one of which is a general law. Similarly, to *predict* the occurrence of some event is to derive a statement describing the event from other statements, at least one of which is a general law. Explanation and prediction are thus *structurally similar*; they have

[29]This point is emphasized by Ernest Nagel in the course of an extended discussion of laws and counterfactuals, in *The Structure of Science* (New York: Harcourt, Brace & World, 1961), pp. 47–78.

[30]In this case, perhaps most importantly to a molecular-chemical theory.

the same form. The only difference between them concerns the time at which the explanation or prediction is made. Other things being equal, if one derives a description of an event prior to its having occurred, the event has been predicted; if the description has been derived after the event has taken place, it has been explained. We explain why a given thread *T* broke, and predict that the same thread *T* will break with reference to the same laws and initial conditions. What varies is simply the time at which we make reference to these laws and conditions.

This alleged symmetry of explanation and prediction deserves a closer look. The first thing to notice is that only structural similarity is claimed. In particular, it is not claimed that on the basis of present data one can make deductive inferences concerning the past as well as the future. The former constitutes *retrodiction,* not explanation. Explanation is always with reference to conditions antecedent to the event to be explained, as is prediction. Retrodiction, on the other hand, involves reference to conditions that post-date the event in question.

An illustration should help to clarify the point. The early nineteenth century French physicist and mathematician LaPlace put forward a celebrated thesis, to the effect that:

> An intelligence knowing all the forces acting in nature at a given instant, as well as the momentary positions of all things in the universe, would be able to comprehend in one single formula the motions of the largest bodies as well as of the lightest atoms in the world, provided that its intellect were sufficiently powerful to subject all data to analysis: to it nothing would be uncertain, the future as well as the past would be present to its eyes.[31]

If we interpret this passage narrowly, and assume both that he is speaking about a closed system and about Newton's laws only, then LaPlace overstates his case in at least two ways. First, the superhuman intelligence would not be able to forecast the future course of events in full detail, given information about the forces acting on and the positions of objects. He would be able to forecast only their future positions, momenta, and so on. He could not, for example, predict the color changes they might undergo. Secondly, even to make the more limited forecast with accuracy, he would need to know in addition how the system of objects under consideration is affected by external forces during its subsequent history.

Nevertheless, LaPlace draws our attention to an important

[31] *Théorie Analytique des Probabilitiés* (1820).

point. Given a physical theory like Newtonian mechanics, one can infer, on the basis of information concerning, for example, present positions and momenta, the positions and momenta of objects at any other time, past as well as future. Which is to say that one can retrodict as well as predict certain types of event. One can do this because the laws of Newtonian mechanics are *deterministic* in a very strong sense; they are reversible as regards their time variables.

Many scientific laws, however, are not deterministic in this same strong sense. Other laws (for example, the statistical laws characteristic of the quantum theory) are essentially indeterministic. If we were to insist that an adequate explanation be a potential retrodiction as well as a potential prediction, then laws that were not strongly deterministic could not be used in explanatory arguments. Some philosophers, impressed by the force of explanations in classical physics, have insisted that explanations be potential retrodictions, and accordingly have restricted the kinds of laws that can be used in explanatory arguments to those of a Newtonian character.[32] But since we require of explanations only that they be potential predictions, there is no need for us to make a similar restriction. Moreover, there seems to be no good reason for making such a restriction. We often accept explanations as intuitively sound that could not be turned into retrodictions.

The second thing to notice about the alleged symmetry of explanation and prediction involves a point already made. When explanation (on the "covering law model") is in terms of statistical laws, it cannot be of individual events. Conversely, one cannot predict the occurrence of individual events on the basis of such laws. But note that, were one to allow that events can be explained inductively, the symmetry of explanation and prediction would not necessarily be undermined. For prediction in such circumstances would also be inductive. There is a break-down in the case of retrodiction: it is possible to retrodict an event deductively and explain it inductively, explain it deductively and retrodict it inductively.[33] Which is to say that laws and statistical generalizations can vary as regards their temporal "direction." But if we do not require of an explanation that it be a potential retrodiction, this complication need not bother us.

[32]Notice how one might be led into taking retrodictions and predictions as symmetrical, and as a result confuse retrodictions with explanations. We have spent some time on the issue because the confusion is easy to make.

[33]A particularly good example of inductive retrodiction is radiocarbon (carbon 14) dating.

There is one more preliminary point that should be made. It concerns the use of the words "determinist" and "deterministic." There are a number of separate "determinist" theses, and only confusion results when this fact is not clearly recognized. Three such theses have already been alluded to in one form or another in the course of our discussion. The "weakest" of these is simply that given certain initial conditions and laws or statistical generalizations, statements describing other events follow, either inductively or deductively. In this sense, the "Humean" account of explanation is "determinist": that events of certain kinds will occur follows (once again, deductively or inductively) from the fact that events of certain other kinds have occurred and that there are laws or statistical generalizations connecting both kinds of events. A slightly "stronger" thesis restricts the "Humean" account to explanations that employ universal laws and not statistical generalizations. Given the premises of an explanatory argument on this thesis, the conclusion follows deductively. That a particular event occurred is necessary relative to other antecedent events and laws connecting it to them. A still "stronger" thesis is that explanations are adequate just in case the laws that figure in them allow equally for retrodiction and prediction, that is, that they are of a Newtonian character. The three "determinist" theses are distinguished, then, on the basis of the different sorts of laws that each thesis says adequate explanations must contain. Note also that these theses have to do with the logical character of explanations. One goes beyond them when he maintains that *in fact* nothing or some things or everything can be explained in one or another of the three ways indicated. Thus, the "thesis of determinism" is often put by saying that "everything in the world is causally determined." But to say this is to make a claim not only about the structure of adequate explanations but about the actual constitution of the world as well.

Now to some difficulties surrounding the so-called symmetry of explanation and prediction.[34] The first is that a prediction is typically thought of as an assertion, for example, "the sun will rise tomorrow," not as an argument. This point does not serve so much to distinguish predictions from explanations, however, as it does to force a modification of our account. For "explanation," too, is often used to characterize assertions, not arguments. "Why did Jones die?" "Because he was poisoned." To say that such assertions

[34]We here draw in part on Israel Scheffler's article, "Explanation, Prediction, and Abstraction," *British Journal for the Philosophy of Science*, VII (1957), 293–309.

are not explanations, or are merely "explanation-sketches," is implausible and unnecessary. Our account can be modified so as to count assertions as predictions and explanations, just in case there is an explanatory or predictive argument in which they figure as premises. The central features of our account of explanation are left untouched by the modification.

Note that this modification, as we described it, has another important consequence. We said that an assertion counts as an explanation, just in case *there exists* an explanatory argument. In other words, the adequacy of a particular explanation does not depend on whether or not the person who advances it can specify the laws and initial conditions on which it depends. All that is required is that the laws and initial conditions that, together with the assertion in question, allow for the derivation of the *explanandum* statement exist. Many scientific explanations are not set out in full detail. Of course, to the extent that we do not believe the needed laws exist, we will not be prone to accept the explanation. And in fact, explanations are commonly advanced and accepted, simply on the basis that we have good reason to think that the laws on which they ostensibly rest exist.

The second apparent difficulty with the alleged symmetry of explanation and prediction is that predictions do not always rest on laws as do explanations. Recall our example of a nonlawlike universal. "All persons sitting on a certain bench in Boston are Irish." This generalization, accidental or not, certainly seems to provide adequate grounds for the prediction that person *P* who has been known to frequent that bench in Boston will be Irish.

In the third place, whereas it is necessary for the *explanandum* of an explanation to be *true,* otherwise there would be nothing to explain, it is possible for a prediction to be false. Indeed, when predictions are used to test theories, this feature of them is very important. For the adequacy of a theory partly depends on its ability to yield true predictions. If a prediction turns out to be false, then the theory receives a measure of disconfirmation.

In the fourth place, a predictive argument is not always a potential explanation. This is especially clear in the case of inductive arguments. On the basis of statistical data, we can often project quite accurately that so-and-so many people will die over a certain holiday weekend, and so on. What is involved is a generalization on past cases. But the statistical data in question do not (at least intuitively) constitute *laws* of any sort, hence cannot be used in the *explanation* of, for example, the size of the holiday traffic toll.

These considerations might lead us to give up claiming that explanation and prediction are symmetrical in every case. Predictions may have the formal characteristics of explanations, but often they do not. There is another objection to be considered, however, which is to the effect that in the case of voluntary action or purposive behavior an explanation is *never* a potential prediction.

To explain some action, it is frequently contended, is to refer to the agent's motive, belief, or attitude. We explain the assassination of the President by giving the motive, for example, a desire for headline fame, for which the assassin acted. Further, where we can explain an action in terms of the motive, the purpose for which it was done, the action is voluntary.

Granting that indicating the motive provides an adequate explanation of the action in question, typically two stances are taken. Some philosophers argue that such explanations are not *causal* explanations. Their reason is that there are no *laws* linking motives to actions. Thus, Hart and Honoré claim that, "The statement that one person did something because, for example, another threatened him, carries no implication or covert assertion that if the circumstances were repeated the same action would follow,"[35] and go on to conclude that motives are not causes of actions. Of course, we have admitted that not all explanations are causal explanations, and that our account is adequate only for the latter.

Other philosophers argue that we do have rough generalizations connecting motives and actions that can, with the development of psychology, be formulated as laws. Thus, there is no reason in principle not to regard motives as causes, or the explanations in which they figure as causal explanations.

There are many complex issues surrounding this dispute. We will have more to say about them shortly, when the general question of teleological explanation is taken up. For the moment, we can indicate a reason for rejecting the suggestion that laws linking motives and actions can be formulated. The point is well made by Donald Davidson:

> . . . generalizations connecting reasons (that is, motives) and action are not, and cannot be sharpened into, the kind of law on the basis of which accurate predictions can reliably be made. If we reflect on the way reasons determine choice, decision, and behavior, it is easy to see why this is so. What emerges, in the *ex post facto* atmosphere

[35]H. L. Hart and A. M. Honoré, *Causation in the Law* (Oxford: University Press, 1959), p. 52.

> of explanation and justification, as the reason frequently was, to the agent at the time of the action, one consideration among many, a reason. Any serious theory for predicting action on the basis of reasons must find a way of evaluating the relative force of various desires and beliefs in the matrix of decision; it cannot take as its starting point the refinement of what is to be expected from a single desire.[36]

After the action has been performed, it can be explained in terms of the motive (reason) for which it was, in fact, performed. But prior to its being performed, the action could not have been predicted solely on the basis of the motive. For as Davidson emphasizes, rarely, if ever, does one motive alone enter the matrix of decision. In fact, decision is called for when there are several motives. Thus, we have a situation where explanation and prediction are clearly not symmetrical. Further, it is claimed that this situation typifies the explanation of human behavior. Such behavior is not to be explained causally. If at all, it is to be explained in ways which make no reference, implicit or explicit, to laws.[37]

This said, it remains an open question whether "motive" explanations are irreducible to "causal" explanations, or whether, if not formulated in terms of "motive," "reason," "desire," and "belief," there are nevertheless laws on the basis of which human (and perhaps purposive animal) behavior can be predicted and explained.

6. Teleological Explanation

To explain some event, one might object, is not at all to subsume it, deductively or inductively, under a generalization, "causal law" or otherwise. Different sorts of consideration are relevant. Two variations on this theme are possible. On the one hand, it can be maintained merely that the present account of explanation does not extend to all kinds of events, to all types of phenomena. In particular, the claim typically goes, it does not extend to explanation in biology, history, and the social sciences, where explanation is in terms of the ends or goals of certain processes. Explanation of this kind is called *teleological.* This, of course, is just to deny that every type of phenomenon can be given a causal explanation. On the other hand, one might want to scrap the account as adequate even for causal explanations. When we come to

[36]"Actions, Reasons, and Causes," *Journal of Philosophy,* LX (1963), 697.

[37]Once again, it is sometimes argued that to admit lawlike explanation is to deny "free will." We will take this issue up in the final chapter.

the section on reduction and explanation, some reasons for making this move will be advanced.

We stipulated at the outset that our interest would be primarily in "why" questions that took causal answers; our central concern in this chapter has been to analyze the concept of *causal* explanation. But there is, it has often been urged, another sort of "why" question and another concept of explanation, at least equally deserving of attention. This sort of "why" question requires an answer in terms of ends, goals, purposes, and functions; the concept involved is that of *teleological* explanation.

One of the most interesting chapters in the history of ideas concerns the attack spearheaded by Descartes in the seventeenth century against the pattern of teleological explanation then current among Aristotelians. This attack had two aspects. On the one hand, Descartes argued that teleological explanations were not genuinely explanatory. The mistaken search for them according to him had, in fact, impeded the development of science. On the other hand, Descartes sketched a program that would, he thought, ultimately provide for the causal explanation of all phenomena. A case held up to illustrate his position was William Harvey's explanation of the circulation of the blood. Harvey provided a perfectly satisfactory causal account of the circulation; to ask what further purposes such circulation served, or why God had designed us thus, was unnecessary and unhelpful.[38] But three footnotes must be added to this chapter in the history of ideas.

In the first place, Descartes equated "causal" with "mechanical." To explain the behavior of an object causally was, for him, to explain it in terms of the motion and interaction of the object's parts. All of nature was construed on the analogy of a vast machine, to be understood and explained as such. The science of mechanics (which is what Descartes thought physics was) comprehended all natural phenomena, including biological phenomena.

But our account of the concept of causal explanation does not commit us to equating "causal" with "mechanical," although we might take our account to fit the case of mechanical explanations very nicely. This point is important, for much of the controversy about teleological explanation has involved contrasting it with mechanical explanation. But our concern is not with the possibility of giving, for example, a mechanical explanation of the behavior of living tissue, but rather with the possibility of giving it a lawlike/

[38]Cf. the *Discourse on Method,* fifth part.

deductive explanation. One could, for example, maintain that there are functional laws, namely, that organisms tend to have characteristics that have survival value. On our account, explanations in which laws of this sort figured would be causal, although certainly not mechanical. To put it another way, the central issue for us is not whether biology is a branch of, or "reducible" to, physics.

In the second place, Descartes argued that human behavior could not be causally, that is, mechanically, explained. It followed from his position, of course, that man could not then be a natural object, but he was willing to accept this consequence. There was, he thought, something fundamentally different about men—the fact that they were essentially minds who *had* bodies—that set them apart. Their behavior, in sharp contrast to that of animals, is purposive, although their bodily movements are to be explained in a mechanical way. One of Descartes' arguments in support of this claim touches on our concept of causal explanation: he asserts that much human behavior cannot be predicted, hence cannot be explained causally.

Arguments like Descartes' have been taken up quite seriously by many recent philosophers. They contend that human behavior is not only not to be explained mechanically, it is not to be explained in a "Humean" way. Whether these same philosophers would be willing to press for the primacy of teleological explanations in regard to such behavior is not always clear.[39] What is clear is that for them explanations in terms of reasons, motives, purposes, attitudes, and beliefs are perfectly satisfactory and not rephraseable in terms of laws and deductive consequences. Some argue that there are no laws under which particular actions may be subsumed; others argue more simply that reference to laws is *inappropriate* to the explanation of human action.

In the third place, Descartes' attack on teleological explanation seemed blunted by the emergence of Darwin's evolutionary theory. With regard to that theory, the question, "Why does that organism have such and such structural or functional characteristics?" appears always to require the answer, "In order to survive," or some such. Evolutionary explanation, to put it a slightly different way, appears

[39]Exception made for Charles Taylor's, *The Explanation of Behavior* (New York: Humanities Press, Inc., 1964), where a trenchant critique of causal-behaviorist theories is coupled with a defense of teleological explanation. For Taylor, however, teleological explanation is explanation that employs "teleological laws," whereas for us what in part distinguishes teleological explanation is the fact that it is nonlawlike.

always to be in terms of the last stages of processes, which in turn are taken to be the ends or goals of those processes, the purposes by reference to which the process is to be explained. Can this type of explanation too be subsumed under our account?

How does teleological explanation differ from causal explanation as we have characterized it? Consider the following pair of questions:

Q.1: Why did rat *A* turn right in the *t*-maze?
Q.2: Why did Harvey choose a pre-medical course?

and their natural answers:

A.1: In order to get food.
A.2: In order to qualify for medical school.

Roughly, these explanations turn on the reference made to the purpose for which the action was done; they are explanations of purposive behavior, and for that reason at least prima facie teleological.[40] There are two other criteria for calling these teleological explanations, and by means of which they can be distinguished from causal explanations. One is a *grammatical*, the other a *temporal*, criterion. According to the first, a teleological explanation is one that contains an expression like "in order to" in the *explanans*. According to the second criterion, a teleological explanation is one whose *explanans* refers to an event later in time than the event it explains. For example, getting to food in a *t*-maze occurs later in time than turning right at the choice point. According to both of these criteria, A.1 and A.2 are teleological explanations. While we have not insisted on the point, it seems reasonable to require of causal explanations both that cause precede effect and that the explanation contain the word "because."

With a few adjustments, however, teleological explanations isolated in accord with these criteria can be accommodated to our account of causal explanation. Notice that to the questions:

Q.1: Why did rat *A* turn right in the *t*-maze?
Q.2: Why did Harvey choose a pre-medical course?

one might also answer:

A.1: Because he was hungry, and he believed that if he turned right he would get food
A.2: Because he wants to go to medical school, and he believes that qualification for medical school is dependent upon taking a pre-medical course.

[40]Contrary to Descartes, we suppose that the rat's behavior is purposive.

These latter two explanations suggest that wherever the expression "in order to" occurs in an explanation it can be replaced by a statement making reference not merely to the object desired, but to the desires and beliefs themselves as causal items in the behavior of the person or organism.[41] Thus we have an apparent explanation of the rat's taking the right turn in a *t*-maze:

> E.1: Rat *A* desires to get the food
> E.2: Rat *A* believes that if he turns right in the *t*-maze he will get the food
> E.3: Whenever something that is desired strongly enough is such that some action is believed to result in getting the desired object, then the action will be performed.

From these statements, which describe, respectively, a set of initial conditions and a law, one might claim that

> E.4: Rat *A* takes the right turn

follows.

To the uninitiated "Humean," these moves might look persuasive. But there is much to object to in this alleged "reduction" of teleological explanations to causal ones.

To begin with, E.4 clearly does *not follow from* E.1–E.3. If, for example, E.1 is to be used together with E.3, it will have to be reformulated to read:

> E.1′: Rat *A* desires to get the food strongly enough.

But what sort of premise in an explanation is this? For how else could we measure the strength of rat *A*'s desire to get food, other than by looking to see whether he in fact goes for the food. But surely this is circular: we attribute a desire to the rat and say it *caused* him to act in a certain way, when our only evidence for attributing the desire and making the causal claim was the fact that he ran for the food. In similar ways, the purported "law" is suspect. For it verges on a truth of logic to say that when someone wants to do something *strongly enough* he will do it. The words "strongly enough" insure against counter-examples.

On the other hand, if we don't insure ourselves against counter-examples in this way, the "law" is obviously false. We all know (perhaps from our own case especially) that when one has a

[41]It might be noted that explanations of actions generally include reference to both beliefs and desires ("pro-attitudes").

desire for something and believes that a certain course of action will satisfy that desire, it certainly does not follow that he will take that course of action. As we mentioned in the previous section of this chapter, desires and beliefs, motives and reasons do not allow us to predict behavior with any degree of accuracy, although they do allow us to explain it ex post facto.

This is the dilemma that our objector poses: Laws formulated in terms of "desire," "belief," and so on, are either trivial or false. Hence, they have no explanatory force, and actions cannot be explained by referring to them. Nevertheless, teleological explanations in these same terms can be given. If teleological explanations are to be "reduced," some other way than that already pointed out will have to be taken.

Actually the objector has two issues going for him that, although not unconnected, must be separated. We have so far run them together. The first echoes a requirement Hume puts on causal explanations, namely, that cause and effect be separately describable. What's at stake here is simply this: If a desire is to be taken as a causal item in the explanation of an action, it must be describable independently of the action in which it eventuates. But take the case at hand. Can the desire of the rat be described in any other way than as a desire to get food? Further, what evidence do we have for, on what basis can we attribute to the rat, the desire except that the rat ran through the maze, turned right, and headed for the food? If desires are not separately describable, they cannot be taken as causes of actions.[42]

Hume's point seems to be that if cause and effect are not separately describable, if, for example, we make use of the description of one in describing the other, then our purported "explanation" will be trivial. To use the vocabulary of the first chapter, the tie between cause and effect cannot be *analytic*, if the first is to be genuinely the cause of the second. But, it has been maintained, the connection between desires and actions (more accurately, between descriptions of desires and descriptions of actions) is of an at least quasi-analytic kind. Desires are always desires of some kind. But to indicate the kind of desires is at the same time to indicate the object or course of action the desire is *for*, or *to*, as we have just seen in the rat case. And this is to say that a description of desires includes as part of its meaning a description of objects or courses of action.

[42]Hume himself seems to have taken desires as causes of actions without, however, saying what a desire is or how it is independently describable.

Thus, though reference to desires serves perfectly well to explain an action, we are not entitled to say that desires are causes of actions. We must distinguish very sharply between causes on the one hand, and beliefs, desires, motives, and reasons on the other.[43]

The other issue that the objector has going for him is that *laws* in terms of beliefs and desires, and so on, cannot be formulated. We have already indicated a reason: Either such "laws" will be trivial or they will be false. Nor, as we also mentioned, can rough generalizations in these same terms be sharpened into laws. Notice that our objector is not claiming that these "mentalistic" terms do not allow the formulation of laws simply because they are mentalistic. He wants only to say that their logic, the rules governing their use, prevents them from figuring in a *causal* explanation of human behavior. Once again, the objector adds that a "motive" explanation, causal or not, is perfectly satisfactory. To look for causes in the explanation of behavior is to mistake men for machines. The behavior of the former is goal-directed, teleological; the latter is determined by antecedent conditions.

Let us hasten to add that there are replies to the objector as well. We will consider two. One, which has received wide prominence among many psychologists and some philosophers, is to argue that "belief," and "desire," and so on, can be analyzed out in other, behavioral terms. A classic example is E. C. Tolman's suggested definition of "expectation":

> When we assert that a rat expects food at location *L*, what we assert is that *if* 1) he is deprived of food, 2) he has been trained on path *P*, 3) he is now put on path *P*, 4) path *P* is not blocked, and 5) there are other paths that lead away from path *P*, one of which points directly to location *L*, *then* he will run down the path which points directly to location *L*.
> When we assert that he does *not* expect food at location *L*, what we assert is that, under the same conditions, he will *not* run down the path which points directly to location *L*.[44]

It is not difficult to imagine how parallel definitions of "belief," "desire," and so on, might be constructed. This program of reconstruing all "mentalistic" terms in a behavioral idiom is sometimes

[43]Cf. Abraham I. Melden, *Free Action* (London: Routledge & Kegan Paul Ltd., 1961), Chaps. 8–9.

[44]E. C. Tolman, B. F. Ritchie, and D. Kalish, "Studies in Spatial Learning, I, Orientation and the Shortcut," *Journal of Experimental Psychology*, XXXV (1946), 15.

called *analytical behaviorism.* It provides for the possibility both of independently describing causes and effects and of formulating laws (again in the behavioral vocabulary), hence, ultimately the possibility of causally explaining human behavior. This program has other, related advantages, not the least of which is that behavior (for example, a rat running down a path) is open to the public, intersubjective observation characteristic of science, whereas expectation, beliefs, and desires are at best inferred.

The response to behaviorism, as a psychological theory and as a philosophical thesis, has been varied and complex. Suffice it to say that in neither guise has it found universal acceptance. For one thing, serious difficulties have arisen in the attempt to reconstrue "mentalistic" terms in the behavioral vocabulary.[45] Tolman's suggested definition of "expectation" illustrates some of these. There are a variety of circumstances in which, given the stated conditions, the rat will not run down the path (for example, a cat suddenly crosses it), and yet in which we surely would not conclude that the rat did not expect food at location *L*. If, on the other hand, we were to insist that Tolman's suggestion be taken as the definition of "expectation," then it should be equally clear that the meaning of this expression had been altered rather radically. Along somewhat the same lines, it seems also that every attempt to dispense with "mentalistic" terms inevitably is driven to employ at least one of them.[46] To rule out the influence of the cat crossing the path on the rat's behavior, we must assume that the rat is motivated solely by the desire for food. Which is to introduce another "mentalistic" term. The regress starts; given the fact that we can multiply without limit circumstances in which the rat, although otherwise "expectant," would fail to run down the path, it appears to be infinite. But perhaps of more importance is the still unanswered objection that a "motive" explanation is perfectly satisfactory. It seems, in fact, uniquely appropriate for the description and consequent explanation of human (if not rodent) action.

The second reply to our objector takes root in a reformulation of the "Humean" account. We said earlier that according to Hume every singular causal claim entails the existence of a general law or regularity. This suggests that to a claim indicating a certain belief and a certain desire as causes of action there must stand a law linking kinds of beliefs and desires to kinds of actions. But we have

[45]Again, cf. Charles Taylor, *The Explanation of Behavior,* Chap. IV.

[46]Cf. R. M. Chisholm, *Perceiving* (Ithaca, N.Y.: Cornell University Press, 1957), Chap. 10.

suggested a reason to dismiss the possibility of such laws. The way out is to propose instead that the law corresponding to the singular causal claim is not necessarily formulated in the same concepts as that claim. In the case at hand for example, the appropriate law might be formulated in neurophysiological terms.

This program differs from behaviorism (as we have construed it) in a crucial respect. It does not maintain that "desire," "belief," and so on, can be analyzed or defined in other terms. It claims only that corresponding to desires and beliefs, there are, for example, states of the central nervous system that are connected in a lawlike way with actions. Desires can be redescribed as such states, and the possibility of neurophysiological laws seems real, although as of this date rather remote. It should also be noticed that this manner of dealing with the objector does not deny the appropriateness of "mentalistic" descriptions and explanations of human actions. It asserts merely that such explanations are ultimately causal, in virtue of the *fact* that desires, and so on, can be correlated or identified with physical states of the body.[47]

An example will perhaps make this kind of position clearer. We want to explain purposive behavior. A thermostat seems to be a paradigm case of purposive behavior. We can explain its behavior teleologically. It activates the furnace *in order to* maintain the room in which it is placed at a given temperature. On the other hand, the action of a thermostat can be described in causal, even mechanical terms. Both descriptions and explanations, it is argued, are descriptions and explanations of one and the same state of affairs. There is no incompatibility between them, although one cannot be analyzed or defined in terms of the other. The "teleological" description of the thermostat contains no "mentalistic" terms; we do not ascribe desires and beliefs to thermostats. But, the proponents of the position argue, there is no important difference between the purposive behavior of thermostats (and, for example, homing rockets) and the purposive behavior of human beings. Both are to be understood on the model of complicated "feed-back" mechanisms. It just happens that a man assimilates and operates on a vastly greater amount of information.[48]

The success of this position hinges, as much as anything else, on how the correlation or identification of desires and beliefs with

[47]See D. M. Armstrong, *A Materialist Theory of the Mind* (London: Routledge & Kegan Paul Ltd., 1968).

[48]See Arturo Rosenbluth, Norbert Wiener, and Julian Bigelow, "Behavior, Purpose and Teleology," *Philosophy of Science*, X (1943), 18–24.

states of, for example, the central nervous system is to be understood. Again, there are numerous variations. What is perhaps most important to realize is that the reducibility or irreducibility of teleological explanations turns at this stage on what is really a metaphysical issue, namely, what is the relation of mind to body. It may be disappointing to be told this. On the other hand, it reminds us that although problems in the philosophy of science arise in rather well-defined contexts, they frequently lead us into extremely difficult metaphysical directions.

There is a last tack one might take with respect to teleological explanations, a tack that appeals more, perhaps, to the scientist than it does to the philosopher. This is to give up "reducing" teleological to causal explanations. One resolves at the outset to give causal explanations and fixes on an appropriate vocabulary. No attempt is made to analyze "desire" in other terms or to identify desire with a physical state. Rather, one just begins with certain notions as primitive, perhaps now not widely used in the description of human behavior, and stakes his case on, for example, his eventual success in controlling and predicting that behavior. This is how many psychologists look at behaviorism. They see it not as "reducing" originally teleological explanations but as "replacing" them.[49]

7. Explanation and Reduction

On the view we have just been considering, there is a kind of phenomenon—purposive behavior—that does not lend itself to explanation along "Humean" lines. A fundamentally different pattern of explanation, which makes reference to motives or reasons and not to laws, is appropriate to it. In this section, we want to discuss still another pattern of explanation. It might be called *reductive* explanation. It remains to be seen whether it is incompatible with, or complimentary to, either of the patterns already discussed.

The connection between the notions of explanation and reduction we have in mind stems from two principles that have been widely held, for the most part implicitly, to determine the adequacy of explanations of a certain type. The first, roughly stated, requires that the properties of wholes be explained in terms of the properties of their parts. It is often referred to as the "principle of micro-reduction." The second principle, again to state it roughly, requires

[49]This position is best exemplified in the writings of B. F. Skinner. See *The Behavior of Organisms* (New York: Appleton-Century-Crofts, 1938), Chap. II.

that the properties of these parts differ from those of the wholes that they are invoked to explain. We will call it the "principle of property-reduction." In a way it serves primarily as an amendment to the first principle. For example, suppose that, in accordance with the principle of micro-reduction, we explained the transparency of water in terms of the molecules and eventually the atoms that compose it. Intuitively, the principle of property-reduction suggests, to say that water is transparent because its components are themselves transparent would be to postpone rather than to provide an adequate explanation of the phenomenon. It is micro-reductive "explanations" of this allegedly trivial kind that the principal of property-reduction is meant to exclude.

This pattern of explanation is quite plausibly taken to be an important source of the persistent search for *atomistic* theories in science, and the accompanying belief that the phenomena have not really been understood until such theories have been found to explain them. It dates, perhaps, from the writings of the Greek "atomist" Democritus. Such is the view of the physicist Werner Heisenberg, who formulates the principle of property-reduction as follows:

> It is impossible to explain . . . qualities of matter except by tracing these back to the behavior of entities which themselves no longer possess these qualities. If atoms are really to explain the origin of color and smell of visible bodies, then they cannot possess properties like color and smell . . . Atomic theory consistently denies the atom any such perceptible properties.[50]

This, according to Heisenberg, is Democritus' basic insight. If the phenomena have not been "reduced" in this way, decomposed into parts that have different properties than those of the wholes whose behavior they explain, then they have not been adequately understood.[51]

A case from physics should help to illustrate this account. Thermal phenomena of various kinds are studied by the science of thermodynamics. The laws of thermodynamics—of which those deal-

[50]Quoted by Norwood R. Hanson, *The Concept of the Positron* (Cambridge: Cambridge University Press, 1963), p. 50.

[51]It is sometimes claimed that to explain something is to "reduce it to the familiar." The difficulty with this suggestion is that "the familiar" and "reduce" (as here used) are extremely elusive notions. To the extent that we do understand them, it seems in fact that historically the development of science has been in the direction of explaining the familiar in terms of the less familiar. In any case, the connection we have in mind between explanation and reduction is of quite another kind.

ing with the impossibility of building perpetual motion machines of assorted types and the Boyle-Charles law are the most famous—are of the functional dependence type. The Boyle-Charles law, for instance, asserts what a functional relationship holds between the pressure, volume, and temperature of an ideal gas, that is, $PV = RT$ (where "R" is some constant). On the "Humean" account of causal explanation, to explain some thermal phenomenon is simply to subsume it, along with specified initial conditions, under one of these laws. But the phenomena of heat can also be studied in terms of the science of statistical mechanics and the kinetic theory of matter. For if we assume that, for example, a gas is composed of molecules having only mechanical properties, and make certain idealizing assumptions, then on this basis we can derive experimental laws that approximate those of thermodynamics and describe the phenomena in question. Explanation in terms of statistical mechanics and the kinetic theory, moreover, conforms very closely to what we have called the reductive pattern of explanation. In the first place, certain macroscopic properties of objects—notably temperature—are reduced to and explained in terms of the component parts—molecules—of those objects. The explanation thus embodies the principle of micro-reduction. In the second place, these parts, as characterized by the micro-theory that introduces them, lack those macroscopic properties of objects that they are invoked to explain. In the example at hand, temperature—the concept of which does not appear among the fundamental assumptions of statistical mechanics or kinetic theory—is explained in terms of, among other things, the positions and momenta of molecules.[52]

We have mentioned that thermodynamics as satisfactorily explains thermal phenomena as does statistical mechanics and kinetic theory on the "Humean" account, insofar as thermodynamics allows descriptions of these phenomena to be subsumed deductively under laws. But many physicists themselves tend to regard only statistical mechanics and kinetic theory as genuinely explanatory, distinguishing it in this respect from other, *phenomenological,*[53] theories (among

[52]Galileo was perhaps the first to give a systematic reductive account of heat, explaining it in terms of the motions of atoms. Galileo also was among the first (in modern times) to link such a reductive account to a distinction between so-called "primary" and "secondary" qualities. The former were "real," the latter merely "apparent," qualities of objects, the essential difference between them being that the latter could be "reduced to" (explained in terms of) the former. Thus, heat (a "secondary" quality) is explained in terms of motion (a "primary" quality).

[53]We might also say "merely descriptive."

which thermodynamics is included).[54] More generally, they tend to distinguish between reductive (for example, molecular and microphysical) theories and others along these same lines. Until we have such theories, it is contended, we do not see "how things really work."

Interestingly enough, a small number of philosopher-scientists have resisted the general tendency to search for atomistic theories, notably Ernst Mach and Pierre Duhem. According to them, this tendency derives not from any empirical considerations (surely atoms are not perceived by the senses) nor from the necessity of satisfying certain logical conditions. It stems rather from an a priori metaphysical view of the world. According to Mach and Duhem, the preference shown among physicists for statistical mechanics and the kinetic theory of matter vis-a-vis thermodynamics is simply a prejudice. But at the same time, those who implicitly abandon the principles of reduction we have listed quite explicitly give up the claim that scientific theories are *explanatory*. They grant that to explain is at least in large part to reduce. Their position is that scientific (as contrasted with "metaphysical") theories do not explain anything; their whole aim is instead to describe, classify, and connect phenomena.[55]

We have restricted our discussion so far to physics. But reductive explanation appears also to characterize other sciences.[56] It is, for example, a common view that to explain the behavior of nations adequately is to frame a theory in terms of individual men, the description of whose actions requires a different vocabulary. Again, it is often maintained among economists that although macroeconomic phenomena like gross national products can be described as such, they can only be explained in the microeconomic terms of price and wage. There is an important difficulty here: although there is a sense in which physical objects, nations, and gross national products are "wholes," and molecules, individual men, and prices

[54]Thus, Enrico Fermi: ". . . statistical mechanics has led to a very satisfactory understanding of the fundamental thermodynamical laws . . . thermodynamical results are generally highly accurate. On the other hand, it is sometimes rather unsatisfactory to obtain results without being able to see in some detail how things really work . . ." *Thermodynamics* (New York: Dover Publications, Inc., 1937), pp. ix–x.

[55]This position is expressed in very strong fashion by Pierce Duhem in the first chapter of his book, *The Aim and Structure of Physical Theory* (New York: Atheneum Publishers, 1962).

[56]Although perhaps the enormous success of physics is enough to account for the effort to extend micro- and property-reductive explanations to other kinds of phenomena.

"parts," it is not at all clear that in each case the whole-part relation is the same. Each case does seem to involve a "reductive" pattern of explanation; how precise a concept of reduction can be made to fit them all is another question.

There is, however, a way in which reductive explanation is more intimately related to physics. It is a long-standing claim, pressed by Descartes among others, that science has a certain unity about it, in that all of its separate branches ultimately reduce to physics. Stated crudely, the claim runs somewhat as follows: Each brand of science has associated with it a particular kind of object. These branches may be arranged in a hierarchical order in terms of the complexity of these objects. At the bottom of the hierarchy stands physics, the branch of science which studies the objects (currently mesons, and others) of which the objects studied by the other branches are ultimately composed.[57] Thus, psychology studies human behavior. But men are composed of cells. Since biology studies cell behavior, it is a more basic science than psychology, in the sense that the explanations it provides are more all-encompassing and profound. But cells are composed of submicroscopic particles. Hence physics, which studies the behavior and simple properties of these particles, is ultimately more explanatory than biology. This theme, as well as atomism, dates from the Greek philosophers; indeed the two themes are probably interrelated. To explain all natural phenomena in terms of the properties of their least parts seems to have been a goal of science for a very long time.

In making these scattered remarks about reductive explanation, we have delayed any direct comparison of it with the "Humean" account. In fact, only when we come to the final chapter, on the limits of science, will the question of their possible incompatibility be taken up. But there is one contrast between them that could be mentioned now. On the "Humean" account, explanation and prediction are taken to be symmetrical. To explain some phenomenon is to be able to have predicted it. For this reason laws play a central role in explanation. But certainly reduction is not a restriction placed on prediction,[58] although we have suggested that it is a requirement that many types of explanations must satisfy. On the

[57]See Paul Oppenheim and Hilary Putnam, "Unity of Science as a Working Hypothesis," in *Minnesota Studies in the Philosophy of Science*, Vol. II, eds. H. Feigl, M. Scriven, and G. Maxwell (Minneapolis: University of Minnesota Press, 1955).

[58]That is, prediction of some event does not require that it be described in microphysical terms.

reductive pattern of explanation, explanation and prediction are not necessarily symmetrical.

Reduction seems, moreover, to have little to do with the notion of prediction—to be independent of it. How then can the reductive account of explanation satisfy one of our original criteria of adequacy, that any explanation indicates how the event in question could have been expected? There appear to be three different answers. The first is that the notion of reduction only serves to supplement the "Humean" account in certain areas. Reductive explanations, the answer runs, have a "Humean" form. They essentially involve deductive subsumption under laws. They therefore allow for prediction. As mentioned, we will have something more to say about this in the last chapter.

A second possible answer is that our criteria of adequacy must be revised. An explanation need not indicate why, given certain antecedent conditions, the event to be explained could have been expected to occur. On the other hand, it might be urged that an explanation must describe the "inner workings" of events.

A third possible answer is suggested by Mach and Duhem. It is to say that while science does enable us to "expect" events, and to this end deductively subsumes them under laws, it does not explain them. Theories are instruments of prediction; they are not explanatory. Explanation does involve reduction, but science involves neither. The difficulty with this very "positivist" answer is obvious. To say that science gives predictions but not explanations is far too counter-intuitive. Science is, as Nagel writes in the passage quoted at the outset of this chapter, the paradigmatic explanatory activity. To say what it is to explain anything is at least in large part to say what it is to give a "scientific" explanation of it. If there are no explanations in science, there are no explanations at all.

For Further Reading

Feigl, H. and G. Maxwell, eds., *Minnesota Studies in the Philosophy of Science*, Vol. III (Minneapolis: University of Minnesota Press, 1962).

Hempel, Carl, *Philosophy of Natural Science*. Englewood Cliffs, N.J.: Prentice-Hall, Inc., 1966.

Nagel, Ernest, *The Structure of Science*. New York: Harcourt, Brace & World, 1961.

Popper, Karl, *The Logic of Scientific Discovery*. London: Hutchinson & Co. (Publishers) Limited, 1959; New York: Basic Books, Inc., 1959; revised ed. paperbound New York: Harper & Row, Publishers.

Rudner, Richard, *Philosophy of Social Science*. Englewood Cliffs, N.J.: Prentice-Hall, Inc., 1967.

Scheffler, Israel. *The Anatomy of Inquiry*. New York: Alfred A. Knopf, Inc., 1963.

CONFIRMATION AND ACCEPTABILITY

4

1. Introduction

We noted in the previous chapter that the "Humean" account of explanation turns on an important distinction between *laws* and *accidental generalizations.* Not every statement of universal form serves to explain the events that it subsumes or "covers"; only lawlike statements do. The distinction suggested was that lawlike statements, but not accidental generalizations, support corresponding counterfactual conditionals to the effect that if some state of affairs were realized, then so would some other state be realized. We also noted that this suggestion does not take us very far. For it quickly runs up against the fact that such conditionals in turn are to be analyzed in terms of the lawlike statements that support *them.* Without a prior grip on the notion of a lawlike statement, our understanding of them is weak at best. But there is a way out. The circle can be broken by independently characterizing lawlike statements as those generalizations that are *confirmed by their positive instances.* Thus we are brought to the twofold problem of confirmation: What constitutes a positive instance of a generalization and under what conditions do positive instances of a generalization confirm a generalization?[1]

[1]More formally the problem is to define the relation which obtains between a singular statement S_1 and its generalization S_2, in those cases in which the former can genuinely be said to confirm the latter. In this respect, our discussion will be somewhat limited: there is no demand that this relation be quantitative (that is, that it indicate the degree to which S_1 supports or confirms S_2). Our interest is solely in so-called "qualitative" confirmation.

We have another reason for being interested in the problem of confirmation. Our interest is not only in distinguishing between lawlike statements and accidental generalizations, but in being able to say how our theories are supported by the evidence we have for them. We want to know, as did Hume, what the grounds are upon which we are entitled to predict the occurrence of future or otherwise unsampled events. This is one form of the *problem of induction:* how are we to distinguish between valid and invalid predictions, that is, between those occasions on which we are justified in going beyond our present evidence and those on which we are not? Our discussion in the previous chapter suggests that those predictions are valid that are made in conformity with lawlike statements; induction is grounded in them. However, if this is the case, we must once again be able to single out the lawlike statements, to specify those generalizations that are confirmed by their positive instances. A justification of inductive practices, as well as of our account of explanation, requires that what we have termed the problem of confirmation be solved.[2]

In asking which generalizations are confirmed by their instances (that is, which generalizations are lawlike statements), we are also asking a more general question; we are asking for an analysis of the relation that obtains between hypothesis and evidence. For those hypotheses are to be taken as laws that the evidence we have supports. In addition to being able to distinguish between generalizations that are confirmable and those that are not, that is, between those that are lawlike and those that are not, we want to be able to say under which conditions the confirmable hypotheses are in fact confirmed. Thus, the problem of confirmation goes beyond a particular account of explanation or induction. Any clarification of the term "science," and especially of the term "empirical science," must say something about the relation between hypothesis and evidence. Indeed, it is frequently taken as the hallmark of the scientific enterprise that its hypotheses are controlled by empirical evidence to a much greater degree than any other intellectual activity. Its theories, on the common view, are confirmed by the facts; hence their widespread, virtually compelled acceptance.

2. Acceptability and Confirmation

Acceptability is, in fact, a broader notion than confirmation. Individual scientific hypotheses are accepted in part as a result of their having been confirmed (for example, by laboratory observations) to a greater or lesser degree, but only in part. Accepta-

[2]This way of putting the problem derives from Nelson Goodman's extremely valuable discussion in *Fact, Fiction, and Forecast*, see especially Chapters I and III.

bility depends on other factors as well. We will mention two: simplicity and theoretical connection. Neither of these factors, it should be added at once, is easy to assess.

Certainly of two hypotheses equally satisfactory in other respects, we habitually choose the simpler. Reasons are not hard to find. The simpler hypothesis is usually more elegant,[3] more convenient to work with, more easily understood, remembered, communicated, and so on. But none of these reasons (or any combination of them) provides an objective criterion of simplicity; all trade on an intuitive and somewhat subjective notion. Therefore, a first difficulty is how to specify such an objective ("nonpsychological") criterion.

A second difficulty is how to justify or explain the fact that we *tend* to regard the simpler of alternative hypotheses as having the better chance of being true.[4] We do not always regard the simplest hypothesis as the one that has the best chance to be true; we sometimes reject hypotheses—usually of the conversational variety—because they are *too* simple ("simple-minded"). However, by and large, we do seem to presume a close connection between confirmation and simplicity, a presumption so deeply embedded that we don't think twice about drawing a smooth curve through a given set of plotted points. Copernicus staked a good deal of his case for a heliocentric theory of the solar system on the same presumption. That it was much simpler than the geocentric alternative was, he argued, evidence in favor of its being the *true* theory.

Taken as a criterion of acceptability, how is simplicity to be measured and how justified or explained? Efforts at justification—in particular, of a connection between simplicity and confirmation—have traditionally taken much the same form as efforts at justifying induction.[5] Some are metaphysical. Thus, some argue for, or some-

[3]Aesthetic considerations often figure prominently in the acceptability of hypotheses. Here we lump such considerations with the issue of simplicity, although it is not always advisable to do so. The mathematically most elegant (or beautiful) theory is not necessarily the simplest (relative, of course, to particular characterizations of elegance and simplicity).

[4]". . . the simpler of two theories is generally regarded not only as the more desirable but also as the more probable." W. V. Quine, "On Simple Theories of a Complex World," first published in 1960 and reprinted in *The Ways of Paradox*, p. 242. Quine lists four causes (if not grounds) for this tendency: wishful thinking, "a perceptual bias that slants the data in favor of simple patterns," "a bias in the experimental criteria of concepts, whereby the simpler of two hypotheses is sometimes opened to confirmation while its alternative is left inaccessible," "a preferential system of scorekeeping which tolerates wider deviations the simpler the hypothesis."

[5]Not surprisingly, for what is being justified is the claim that the simpler the hypothesis the more probable that it is true.

times simply assert, a "principle of the uniformity of nature" or a "principle of the simplicity of nature," typically on the philosophical ground that the scientific enterprise itself "presupposes" such principles.[6] Others are methodological. Some proceed *as though* induction worked, or nature were simple, and justify the procedure by the significance of the results gained. To make a very long story short, these moves have increasingly fallen out of favor. The metaphysical efforts are usually thought to beg the question: the principles that are intended to justify induction and simplicity themselves require at least as much justification as the principles of induction and simplicity. The methodological alternative, many feel, misses the point of the question: the success of the scientific enterprise does not by itself show why certain preliminary assumptions are indispensable.

There are other types of arguments on behalf of simplicity. One was especially popular in the earlier part of this century. It is that scientific hypotheses are no more than summaries of recorded observations; their purpose is simply to abridge the catalogue of such observations. On this view, simplicity is an important criterion of acceptability. If science is merely a way of classifying and condensing data, then the most economic hypotheses are to be sought. The difficulty is that this conception of science is a dubious one. Hypotheses are not merely summaries of recorded observations; they go beyond what has been observed to allow prediction of future cases, to guide our expectations in a number of different ways. In fact, hypotheses that describe and abridge the same set of facts can lead us to make very different predictions.

A somewhat extended example should help to clarify the point. It will also serve to introduce certain difficulties in the way of objectively characterizing simplicity.[7] On the basis of certain measurements, we are led to suspect that one quantitative characteristic, v, of a given physical system is functionally dependent on another such characteristic, u—as, on the basis of Boyle's Law,

[6]Leibniz seems to have taken simplicity as the first principle of science. He thought that science "presupposed" simplicity in the sense that from it all the particular laws could be derived. He argued, for example, that Snell's Law ("A ray of light traveling obliquely from one optical medium into another is refracted at the separating surface in such a way that the ratio $\sin \alpha / \sin \beta$ of the sines of the angles of incidence and of refraction is a constant for any pair of media") describes the *least path* of a light ray through a pair of media, hence follows from the principle of simplicity. Needless to say, the "deduction" is suspect.

[7]The example is Carl Hempel's, *Philosophy of Natural Science* (Englewood Cliffs: Prentice-Hall, Inc., 1966), pp. 40ff. We follow his account of it very closely.

temperature is a function of pressure and volume. Our task is to formulate a law that expresses this law in a mathematically precise way—as, once again, does Boyle's Law: $PV = RT$. The measurements made reveal that in those cases in which u had one of the values 0, 1, 2, or 3, the associated values of v were 2, 3, 4, and 5, respectively. Given this information only, an infinite number of hypotheses can be formulated, all of which fit the data. For example:

H.1: $v = u^4 - 6u^3 + 11u^2 - 5u + 2$
H.2: $v = u^5 - 4u^4 - u^3 + 16u^2 - 11u + 2$
H.3: $v = u + 2$
.
.
.

The difficulty for the view that scientific hypotheses are "economic expedients for thought" is that the three hypotheses in question vary radically in their predictions concerning unsampled cases. For instance, when $u = 4$, H.1, H.2, and H.3 predict as values for v, 150, 30, and 6, respectively. Since these hypotheses do vary so radically in their predictions, the grounds for accepting any one of them over the others cannot be that it more successfully abridges the data.

Which one of these hypotheses does most successfully abridge the data; which *is* the simplest hypothesis? Intuitively, perhaps, H.3, but only against a certain background. If we take the order of the polynomial by which all the functions are expressed as an index of complexity, then H.3 is the simplest of the hypotheses. Thus, on a system of rectangular coordinates, H.3 graphs as a straight line, whereas H.1 and H.2 are represented by "more complicated" curves. But if we change the background in certain ways, the ranking of hypotheses in terms of their comparative simplicity changes also. For example, if we represent them in polar coordinates, with u as the direction angle and v as the radius vector, then H.3 determines a spiral, whereas an hypothesis represented by an intuitively "simpler" straight line would be more complicated, for example, the polar equation of the straight line has the form $v \cos (u - a) = p$.

There are a number of different lessons to be learned from this example, some of which have already been hinted at. The first is that a measurement of simplicity is relative to a certain, often mathematical background. Given a particular background, the simplicity of competing hypotheses can be assessed. In the absence of such a shared background, they cannot be assessed except in the subjective way mentioned earlier. The second lesson is that the choice of a particular background seems to be somewhat arbitrary, that background itself being relative to the purposes at hand. But if this is the case, then what entitles us to say either that one hypothesis—say H.3—among the alternatives is "the simplest" or, perhaps more importantly, that it has the best chance of being true?[8]

This point can be generalized in the case of theories. It is sometimes suggested that the simpler of two theories is the one that has the fewer basic assumptions, or fewer basic concepts, or assumes the existence of less objects, or has fewer primitive predicates. But it is extremely difficult to count assumptions, concepts, objects, predicates—there is, for example, no simple correspondence between these and the symbols in terms of which the theory is formalized—and in any case, theories break up the world in different ways, so the assumptions, concepts, objects, and predicates primitive in one are not necessarily those that are basic and primitive in another. Unless embedded within the context of still a third theory that embraces them, there is little hope of assessing the simplicity of any two competing theories. Simplicity seems to be largely an intra-theoretic matter. It is not just any two hypotheses, but rather two hypotheses within a common framework that stand a chance of being compared in terms of their relative simplicity. To put it a slightly different way, the claim that the fundamental laws of nature are simple first depends on the way a particular theory or constellation of theories structures nature. Nature, we might say, is not simple simpliciter.

Theoretical connection is also a factor in the acceptability of hypotheses and theories. We indicated in the previous chapter on explanation that some statements are accepted simply because they follow from other already well confirmed or otherwise accepted laws and theories. For instance, we can suppose that copper at —270° C is a good conductor, not because this result has been confirmed—presumably it hasn't—but because it follows from the fact that all copper is a good conductor. Other statements that already have some measure of inductive support are made more acceptable if they can be related to an existing body of theory. For example, if

[8]Existence of a somewhat common mathematical background allowed Copernicus to argue for the greater simplicity of his theory as against Ptolemy's. But there are ways of recasting the theories mathematically such that the former theory proves to be more complex than the latter. Even given a particular background, moreover, it is still necessary to pick a certain feature of it, e.g., the order of the polynomials, as an index of complexity.

an empirical generalization can be shown to be just a special case of some established law or theoretical principle, its acceptability is very much strengthened.

On the other hand, conflict with established laws or theories serves to diminish the acceptability of an hypothesis, particularly some sort of "low-level" empirical generalization or individual assertion, even in the face of favorable evidence.[9]

There comes a point, of course, when the evidence for the generalization or individual assertion becomes so strong that the laws or theories with which it conflicts are overthrown. But it is as difficult to characterize the point at which laws or theories are disconfirmed, as it is to characterize the point at which they are confirmed. To state the point in the form of a slogan: There are no rejection procedures of standard sorts for theories. One may reject a theory in the face of unfavorable evidence, or one may reject the evidence—especially if the theory is well established on independent grounds, or one may make any number of intermediate adjustments to either theory or evidence or both, for a variety of reasons.

A case often cited from the history of physics should help to illustrate the point. In the middle of the nineteenth century, astronomers were faced by the fact that irregularities in the motion of the planet Uranus could not be explained à la Newton in terms of the gravitational attraction of the other known planets. Three separate tacks were taken. One was to say that the Newtonian theory had been disconfirmed; alternative theories would have to be sought. A second tack, prominently advanced by the Royal Astronomer, was to say that the range of the Newtonian theory, that is, the inverse square law, had to be restricted; the theory satisfactorily explained things out to a distance just short of Uranus, and past that point the behavior of the planets was described by another law—a little

[9]Hume went so far in his celebrated "Essay on Miracles" as to say: "Suppose that the historians who treat of England should agree that, on the first of January 1600, Queen Elizabeth died; that both before and after her death she was seen by her physicians and the whole court, as is usual with persons of her rank; that her successor was acknowledged and proclaimed by the parliament; and that, after being interred a month, she again appeared, resumed the throne, and governed England for three years. I must confess that I should be surprised at the concurrence of so many odd circumstances, but should not have the least inclination to believe so miraculous an event." Hume's point is that even in the face of eyewitness reports we would not accept such a serious exception to the established laws of nature.

mathematical ingenuity sufficed to formulate a law that accounted for Uranus' irregularities.[10] The third tack, taken independently by Adams and Leverrier, was to postulate the existence of an undiscovered planet with a gravitational attraction that would explain Uranus' irregularities. They were spectacularly vindicated a short time later when such a planet, Neptune, was in fact discovered.[11] Prior to that vindication, the observed irregularities did not dictate a unique response. There were many different options to take with respect to both theory and evidence.

We will want to come back to this point about the options available when theory and evidence conflict. Some believe that there lies in it the seeds to a possible "solution" to the problem of confirmation (and induction). An additional comment on it needs to be made now. Of the three different tacks taken, rejection of Newton's theory was the least widespread. There was no doubt about the irregularities, nor the fact that the unamended inverse square law could not deal with them. In the absence of an alternative theory that can explain the phenomena in at least as satisfactory a way as the prevailing theory, we hesitate to abandon the prevailing theory. Rejection of laws or theories typically takes place only within the context of competing alternatives.

There is a third way in which theoretical connection is a factor in the acceptability of hypotheses. An empirical generalization, no matter how well confirmed—that is, no matter how many positive instances of it are piled up—will not be accepted as a law, or more generally, as having any kind of systematic or theoretical import, if a well-established theory permits exceptions to it. In the previous chapter, for instance, we concluded that "Everyone sitting on a certain bench in Boston is Irish" and "All unicorns are fleet of foot" were not laws, simply because nothing in current scientific theory precludes that anyone sitting on a certain bench in Boston

[10]This tack may look very *ad hoc*, but it is taken frequently. In the face of strong disconfirming evidence, a well established theory is typically restricted so that its evidence class no longer contains disconfirming instances, and not simply rejected out of hand. Thus, Newton's theory—given the kinds of results for celestial distances which led to Einstein's theory of relativity—is typically held to be valid for more or less terrestrial distances. A parallel situation often occurs in mathematics: if a contradiction is found in a theory, the theory is not immediately dismissed. Rather, the contradiction is "cordoned off" so as not to infect other parts of the theory. For example, the procedure of cordoning off the difficulty is the one that was followed when the paradoxes were discovered in set theory.

[11]The discovery of Neptune was facilitated in large measure by the fact that the astronomers now knew exactly where to look.

is not Irish nor that unicorns are not fleet of foot. Confirmation is a criterion of acceptability via the criterion of theoretical connection just as, we suggested earlier, it is via the criterion of simplicity. All three criteria are, in fact, inseparable. Failure to realize this interconnection, as we shall soon see, has led to problems.

At this point, more has to be said about confirmation itself. Much about the concept is unproblematic. Thus, we consider a hypothesis to be the better confirmed the more our evidence for it, the more varied and precise the evidence, the greater the extent to which factors not previously known or explained can be brought under it. The difficulty is to say what constitutes the evidence, the facts that it explains.

Our concern has been principally with laws that we took to have the form of *generalized* conditionals. For all objects x, if x has the property A, then it has the property B. A little symbolism will be helpful in what follows. We can represent a generalized conditional as follows:

$$(x)\ (Ax \rightarrow Bx).$$

In this formula "(x)" is short for "for all objects x," and "$\rightarrow$" is our symbol for "if____then. . . ." The formula can also be read more idiomatically as "all A are B." Laws are confirmed, we also said, by their positive instances. Thus, if we find an object a which has property A and also has property B, we have a positive, hence confirming, instance of our generalized conditional. Conversely, if an object a has property A but does not have property B, it is a negative, hence disconfirming, instance of the same conditional. An object that does not have property A is neutral as regards the hypothesis, whether or not it has property B. Positive, negative, and neutral instances of a generalization are *singular* statements because they purport to be about one and only one object rather than about many, if any, objects.

One might object that not all scientific hypotheses have the form of generalized conditionals. For our purposes this does not much matter; the simplification is deliberate, and does not avoid any of the essential difficulties. The important thing is to see how intuitively plausible this simple picture of confirmation is. For example, how do we confirm the statement that "Any body subject to no external forces maintains a constant velocity"? Simply count up the positive instances, that is, the cases in which particular bodies subject to no external forces maintain a constant velocity. Nothing seems more obvious. Positive statements confirm their generalizations; negative statements disconfirm their generalizations. However, profound difficulties lurk just under the surface of this deceptively simple account of confirmation.

3. The "Paradoxes" of Confirmation

These difficulties can perhaps best be made clear in connection with two so-called "paradoxes" of confirmation. Both focus on the notion of a *positive instance* of the hypothesis it confirms.

The first formulated by Carl Hempel,[12] is generally known as the "raven paradox." Suppose we want to confirm the following hypothesis H:

All ravens are black.

In symbols this is $(x)\,(Rx \rightarrow Bx)$. On the present view, we look around for individual cases of black ravens; they constitute appropriate positive instances of the generalization. But H is logically equivalent—by an elementary transformation of classical logic called *Transposition*—to H':

All nonblack things are nonravens.

In symbols this is $(x)\,(-Bx \rightarrow -Rx)$; where "—" is short for "not" (or "non"), that is, a particular thing that is not black and is not a raven is a positive instance of, and hence confirms, this generalization. But if we require—as seems entirely reasonable—that instances that confirm a hypothesis confirm all hypotheses logically equivalent to it (the "equivalence condition"), we are left with the paradoxical result that anything that is not black and is not raven—for example, the Washington Monument—also confirms the hypothesis that all ravens are black.

Matters can be made worse. H is logically equivalent not only to H' but also to H'':

If anything is a raven or is not a raven, then if it's a raven it's black.

In symbols this is $(x)\,((Rx\ v\ -Rx) \rightarrow (-Rx\ v\ Bx))$, where "$v$" abbreviates "or."

This hypothesis, H'', is confirmed, via its positive instances, by nonravens, and also by black objects. Since H'' is logically equivalent to H, nonravens and black objects—as well as black ravens—confirm the hypothesis that all ravens are black.

[12]We have already referred to the importance of Goodman's *Fact, Fiction, and Forecast* in discussions of confirmation. Reference should now also be made to Hempel's papers, "A Purely Syntactical Definition of Confirmation," *Journal of Symbolic Logic*, VIII (1943), 122–43, and "Studies in the Logic of Confirmation," *Mind*, LIV (1945), 1–26, and 97–121, where many of these issues were first raised.

The feeling of discomfiture that these examples provoke is well expressed in Goodman's phrase, "The prospect of being able to investigate ornithological theories without going out in the rain is so attractive that we know there must be a catch in it." Yet we have been led to this implausible result—that just about anything confirms the hypothesis that all ravens are black—by very natural and intuitive assumptions. The entire situation stands in need of reexamination.

Many different attempts have been made to resolve the "raven paradox."[13] We will consider two that contrast rather sharply with each other. The first is Hempel's own.[14] According to him, the trouble lies neither in our original notion of a positive instance nor in the equivalence condition, but in our thinking that the result to which they lead is implausible. In his words, "The impression of a paradoxical situation is not objectively founded; it is a psychological illusion."[15]

In Hempel's view, this "illusion" has two sources. One is our mistaken tendency to think that generalizations of the form "All *A*'s are *B*'s" are *about A*'s only, that *A*'s constitute their subject class. When this form of generalization is paraphrased into a language more formal and perspicuous than English—in particular, into the symbolism of the logic of quantifiers—it is clear that such generalizations are about all objects whatsoever in some specified domain of discourse. For example, the generalization that "all ravens are black" asserts that if *anything* is a raven it is also black. The class of instances of this generalization, positive and negative, is not restricted to ravens. We are misled, perhaps by English grammar, into thinking otherwise, and hence into (mistakenly) rejecting nonblack nonravens, black nonravens, and so on, as perfectly good instances of this same generalization.

The other source of the "illusion" that our notion of a positive instance leads to paradoxical consequences is the frequent and illicit intrusion of additional information. What leads us to rule out, for example, the Washington Monument, as the subject of an acceptable positive instance of the raven generalization is the fact that we already know of the Washington Monument that it is neither a raven nor black. Hence, it can provide us with no "new" evidence for the generalization; it cannot provide added support for it. But if we referred to some test object as "*a*" and characterized it no

[13]There is a helpful discussion of many of these in Part III of Israel Scheffler's *The Anatomy of Inquiry*.

[14]"Studies in the Logic of Confirmation," *op. cit.*, pp. 1–26.

[15]*Ibid.*, p. 18.

further, then the discovery that it was neither a raven nor black would confirm the generalization, even though in fact the test object happened to be, for example, the Washington Monument. If we assume nothing in advance about objects tested in connection with some hypothesis, if we do not allow any additional information we have about them to intrude, then the "paradox" that nonblack nonravens confirm the generalization that all ravens are black "vanishes," to use Hempel's expression. Only when such information (illegitimately) intrudes are the consequences to which our notion of a positive instance and the equivalence condition lead taken to be paradoxical and implausible.

Two features of Hempel's account must be emphasized. The first has already been mentioned; Hempel retains the original notion of a positive instance and the equivalence condition, electing instead to explain away their apparent conflict with other intuitions we might have about confirmation. The second thing to emphasize is that Hempel's is a "formal" account of the notion of an instance, and hence of confirmation. By this is meant simply that confirming instances of a generalization can be determined solely on the basis of the form of their description. For example, any observation report of the form "Ra and Ba" thereby counts as an instance of the hypothesis "$(x)\ (Rx \rightarrow Bx)$." Similarly, two hypotheses are judged to be *equivalent* if they are logically equivalent. To put these points in a slightly different way, what determines the instances, confirming and disconfirming, of a generalization is the language—vocabulary and grammar—in which the generalization is formulated.

The second response to the "raven paradox" we want to consider here finds fault with both these features of Hempel's account. This response sides with, and so does not attempt to explain away, those of our intuitions about confirmation that conflict with the original notion of a positive instance and the equivalence condition, contending that neither can be characterized in a formal way. Thus, this approach has two significant features of its own: on the one hand, it argues for a rejection of solely formal accounts of confirmation and, on the other, it proposes other criteria by which to characterize the notion of a confirming or disconfirming instance of a generalization.

The argument for a rejection of solely formal accounts is simple: the implausible consequences to which they lead, for example, the "raven paradox," cannot be explained away, à la Hempel, as somehow "illusory." Rather, it is the formal account itself that must give way.

The main line of attack might be based on what is taken to be a more adequate description of the scientific enterprise. Certain points that emerge from such a description are crucial. We can list them briefly. In the first place, the scientist typically does not count just anything as a confirming or disconfirming instance of the hypothesis that he tests. The class of potential instances is vastly more delimited. Usually only a very few test cases are considered.[16] In the second place, information of the kind Hempel urges should be excluded constantly intrudes, and it is at least controversial whether or not it intrudes illegitimately. It is often such information, in fact, that allows us to rule out in advance whole classes of cases as potential instances. To suggest, for example, as Hempel does that there are descriptions under which a piece of ice would count as evidence for the hypothesis "All sodium salts burn yellow" seems counter-intuitive at best. In the third place, isolated hypotheses are rarely if ever put to direct experimental test in the way indicated. The testing of hypotheses, and the characterizing of evidence for or against them, is a great deal more complicated conceptually.

This point is not as clear as it might be, but it can be spelled out by way of examples. We can begin with the raven generalization. Confirming or disconfirming this generalization, one might want to say, is not a simple matter of discovering positive and negative instances.[17] Suppose we were to find a white raven, that is, a bird alike in every respect to ravens but white. Is the next move to say that our generalization has been at least partially disconfirmed? Not necessarily. If, for example, we believe that our white raven has been exposed to radiation, and that X-ray radiation affects the color of birds, we might be led to restrict our generalization: all unradiated ravens are black. Again we might reject this one bird as a pathological case, or we might even be led to reject our criteria for ravenhood and maintain that radiated ravens or birds that look like ravens but are white are no longer ravens. The particular move we make is not arbitrary; it is dictated by a variety of different considerations and circumstances. The point is that the finding of a "white raven" does not by itself dictate a response. By the same token, whether or not the case of a white raven is to count as an instance,

[16]It is the fact that test cases are so circumscribed that allows scientists, typically, to consider a hypothesis confirmed on the basis of a few tests. There seems to be very little of the piling-up-of-positive-instances picture often associated with the notion of confirmation.

[17]This claim is true even if we restrict ourselves to the class of ravens.

confirming or disconfirming, of the hypothesis depends on many factors other than the form of its description.

There has, in fact, been a great deal of controversy in the history of science about the way in which particular cases should be treated. We have already mentioned one: Do irregularities in the orbit of the planet Uranus disconfirm Newton's original theory, or do they necessitate restricting the scope of the theory, and so on? A priori, it is impossible to say. Further, it was extremely difficult to determine, even in the case of the laws of motion, what were to count as confirming instances. A great many theoretical/experimental questions hung in the balance. Certain experiments that were thought to disconfirm it were ruled out as eccentric or reinterpreted —for example, as Lorentz and Fitzgerald reinterpreted the results of the Michelson-Morley experiment so as to square them with Newtonian theory, and so on.

The preceding remarks, of course, do *not* suggest that there is no such thing as an instance, or a positive instance, of a hypothesis, or that generalizations are never confirmed or disconfirmed, or that the facts do not control the kind of speculation scientists engage in. What is suggested is that what counts as an instance is not determined by *form* alone. A variety of different considerations are relevant, most of them tied to the theoretical framework in which the particular hypothesis at stake is embedded.[18]

There is another, closely connected issue here. It has to do with the use of the term "theory." We have used it in a fairly free-swinging way, without apology or elaboration. But something more has to be said. "Theory," like "cause" and "explanation" and "law" (and "confirmation" for that matter), is not the clearest of scientific terms. In particular, "theory" is sometimes characterized in a purely formal way, for example, as consisting of a set of "uninterpreted" sentences in the sense that the meaning and subject matter of these sentences is left undetermined. But if it is characterized in this way, then all our remarks about the centrality of theoretical considerations in determining what counts as a positive instance of a hypothesis do not tell against a formal account of confirmation.

On the other hand, we have suggested a way in which the use of "theory" differs from the purely formal account, especially during our consideration of the second account of the "raven

[18]A wealth of examples illustrating the points made in these last paragraphs can be found in Thomas S. Kuhn's important book, *The Structure of Scientific Revolutions* (Chicago: University of Chicago Press, 1962).

paradox." It is that a theory involves essentially specification of the classes of objects to which it applies. What this comes to, in the case of the raven generalization, is easily seen. In addition to the universal conditional, "$(x)\ (Rx \rightarrow Bx)$," the hypothesis includes, to use Scheffler's term, a "field of application"—in this case the class of ravens. Even this limited characterization of a theory is enough to undermine the "raven paradox." For example, nonblack objects are no longer counted as instances of, hence as confirming, the hypothesis that all ravens are black. Perhaps this result alone is persuasive evidence that a purely formal characterization of "theory" is not sufficient.[19]

Many of the points we have been discussing in this section also arise in the second "paradox" of confirmation, the famous "grue paradox" formulated by Goodman.[20] This paradox may be explained as follows. We are asked to suppose that all emeralds examined before a certain time *t* are green. Given our original notion of confirmation via positive instances, at time *t* our observations support or confirm the hypothesis that all emeralds are green. "Our evidence statements assert that emerald *a* is green, that emerald *b* is green, and so on: and each confirms the general hypothesis that all emeralds are green. So far, so good."

At this point, Goodman introduces a new predicate, "grue," which "applies to all things examined before *t* just in case they are green but to other things just in case they are blue." Now consider the two hypotheses:

H.1: All emeralds are green.
H.2: All emeralds are grue.

It should be clear from the way in which "grue" was introduced that at time *t* all the evidence for H.1 is also evidence for H.2, and vice versa. They are equally well confirmed; for at time *t* the two hypotheses have the same positive instances. However, this is paradoxical. In the first place, although we have been forced to say that

[19]An important qualification must be made at this point. It is possible for the syntax of a theory to reflect the "field of application." If we do this, then the above remarks do not necessarily undermine the formal account of theories and confirmation. Most standard "formal accounts" do not reflect the "field of application" of a theory, for example, by characterizing the predicates which occur in the vocabulary of a theory. But there is at least one account of quantification theory in logic which does. Peter Geach argues, on grounds quite independent of ours, that classical quantification theory must be amended—in the interest of capturing the logic of ordinary arguments—to include a specification of predicates. Cf. his *Reference and Generality* (Ithaca: Cornell University Press, 1962).

[20]*Op. cit.*, pp. 73ff.

they are equally well confirmed, they imply incompatible predictions about emeralds subsequently examined. The fact that all the emeralds examined so far have been green—hence also grue—seems not in the least to support the prediction that the next emerald examined will be grue, although it does seem to support the prediction that it will be green. In the second place, "grue" is a totally arbitrary predicate; we have no more reason for thinking that emeralds examined after time *t* will be blue than we have for thinking they will be red. So we have no more reason for asserting "All emeralds are grue" than we have for asserting "All emeralds are gred." We can cook up any number of "grue" type predicates. All will be true of emeralds to the same extent that "green" is; for the generalizations in which they figure are supported by precisely the same evidence. The evidence confirms just about any assertion we wish to make about emeralds, but as in the case of the "raven paradox," this result is, in Goodman's words, "intolerable." In other words, Goodman's story about "grue" conflicts with a variety of entrenched intuitions about confirmation that we have.

We said earlier that *lawlike statements* and *accidental generalizations* are to be distinguished by virtue of the fact that only the former are confirmed by their positive instances, but in light of the "grue paradox," this criterion seems worthless. If any series of observations confirms just about any arbitrary generalization, there is no way to distinguish between lawlike and nonlawlike statements in this respect. Intuitively, H.1 and H.2 are not both lawlike—at the very least, H.2 is not—but our account of confirmation via positive instances does not allow us to distinguish between them or to reject either as merely accidental; they are equally well confirmed by their instances.

Once again, a great number of interesting suggestions for dealing with the "paradox" have been advanced. In one way or another, these suggestions attempt to show how H.1 or H.2 are not equally well confirmed or, equivalently, that "grue" is not a *projectible* predicate in the sense that regularity in grueness does not confirm the prediction (or "projection") of future cases. We want a way of distinguishing between lawlike and nonlawlike statements; hence a way of distinguishing between projectible and nonprojectible predicates. Projectible predicates are inductive with respect to their subjects, that is, the generalizations in which they figure are confirmed by their instances, are lawlike. The problem of confirmation comes to rest here.

Goodman's own somewhat tentative solution to the problem

of distinguishing between projectible and nonprojectible predicates proceeds in terms of a notion of *entrenchment.* We have already seen that we cannot distinguish between

H.1: All emeralds are green.
H.2: All emeralds are grue.

on the basis of their form. Both hypotheses have the form of universal generalizations. Further, since "grue" and "green" are interdefinable, there can be no semantic contrast between them either.[21] They differ, if at all, with respect to some nonformal, nonsemantic characteristic. This characteristic, Goodman suggests, has to do with the use we have in fact made of "green" and "grue" in formulating and testing hypotheses. The solution to the problem of distinguishing between projectible and nonprojectible predicates, more accurately between more or less projectible predicates,

> . . . is that we must consult the record of past projections of the two predicates. Plainly "green," as a veteran of earlier and many more projections than "grue," has the more impressive biography. The predicate "green," we may say, is much better *entrenched* than the predicate "grue."[22]

It is, in other words, the information we have concerning the past histories of predicates that allows us to contrast them, to say that one predicate is more projectible than another because it—or a predicate coextensive with it[23]—is better entrenched than another.

It is important to realize that "entrenchment" is not to be equated with "familiarity." Goodman's suggestion does not rule out relatively unfamiliar predicates like "conducts electricity" and "is radioactive," to use his examples, because they are unfamiliar or admit familiar predicates like "is talkative" just because they are familiar. Entrenchment is a function not of frequency of use but rather of frequency of projection. New predicates that have not previously been used or projected are not ruled out either, just because they are new. Such predicates may be admitted because they

[21]An emerald is grue just in case it is green before time *t* and otherwise blue. Now introduce the predicate "bleen": an emerald is bleen just in case it is blue before time *t* and otherwise green. We can then say that an emerald is green just in case it is grue prior to *t* and otherwise bleen. Notice as a result of this last characterization that we are precluded from saying without further apology that "grue" has a temporal component while "green" does not.

[22]From *Fact, Fiction, and Forecast,* page 95, by Nelson Goodman, copyright © 1965 by The Bobbs-Merrill Company, Inc., reprinted by permission of the publishers.

[23]Two predicates are coextensive when they pick out the same class of objects, for example, "man" and "rational animal" are coextensive predicates.

are coextensive with already well-entrenched predicates, for example, or because particular projections of them do not conflict with projections of much better entrenched predicates. "Grue" conflicts with "green," but it is not easy to see what better entrenched predicate conflicts with "conducts electricity."

At the same time, entrenchment and degree of entrenchment are not very precise notions. As a result, it is somewhat difficult to sort out genuine from merely apparent objections to Goodman's proposal. But, bearing this in mind, there seem to be at least two major problems with it.

On the one hand, Goodman's criterion seems to be too weak. "Green" is more projectible than "grue" because it is better entrenched. No further reason is given why "green" should in fact have been projected earlier and more often. That "green" and "blue" were projected rather than "grue" and "bleen" was a matter of chance. But this is counter-intuitive. "Green" and "blue" were projected because they were clearly more projectible to begin with. Goodman replies that "the judgment of projectibility has derived from the habitual projection, rather than the habitual projection from the judgment of projectibility." What must be done to show that Goodman is wrong is to advance other sorts of reasons why "green" and not "grue" is projectible per se, and not simply that it has in fact been more often projected. The second suggested solution to the "grue paradox" attempts to do just this. We will turn to it in a moment.

On the other hand, Goodman's criterion seems to be too strong. It rules out the possibility of certain kinds of scientific change. Consider, for example, the predicates "has a mass" and "has a weight" from a seventeenth century point of view. From that point of view, "has a weight" is much better entrenched than "has a mass" and hence is more projectible. It follows that we are to frame and test our generalizations about bodies with respect to their weights rather than with respect to their masses. How, then, are we to explain the fact that, beginning in the seventeenth century, "mass" came to be projected instead of "weight," in fact in the course of time to replace it completely as a fundamental physical magnitude? It seems that we cannot, given Goodman's proposal. But it is a recurrent feature of the history of science that shifts from a better entrenched to a less well-entrenched predicate have taken place and a criterion of projectibility must take account of this fact.

A second suggested solution exploits both of these difficulties with Goodman's proposal. From another direction, the second suggestion relates to and enlarges upon the resolution of the "raven paradox" already proposed.[24] It turns on the claim that the notion of an emerald is involved in the lawlikeness of H.1 and H.2. The projectibility of predicates is in large measure a function of the type of thing of which they are predicated. On a very general level, for example, emeralds are physical objects. As such, they can undergo changes while still remaining emeralds, although these changes must be of definite kinds and never uncaused.[25] But the notion of change, and ultimately the concept of causality, is not geared to "grue" type predicates. Something that *stays grue* changes. Given the fact that emeralds are physical objects, and that physical objects are to be conceived in roughly Aristotle's way, then "grue" type predicates are not projectible with respect to them.

We have ruled "grue" type predicates out on very general grounds, but the claim that projectibility is in large measure a function of the type of thing of which predicates are projected has more specific application. Typically, objects are characterized by the particular theories in which they figure, and hence the range of inductive predicates that are inductive with respect to them is determined. That emeralds are physical objects is enough to undermine the lawlikeness of "All emeralds are grue." However, we know a great deal more about them, even on very primitive theories, and this knowledge allows us to narrow the range much further still. To put the point in a very strong way, if you don't know what predicates are projectible with respect to emeralds, you don't know what emeralds are. We do know what emeralds are, relative to some theory; so we do know what predicates are projectible with respect to them.

For instance, given a certain formulation of the atomic theory of matter, there are some things one can and cannot find out about either atoms or matter. One cannot even ask what color they are, or how they taste, or smell. On the other hand, one can ask questions about their position, velocity, and mass. "Red" is not projectible with respect to atoms; "has a mass of *n* milligrams" is. Or, to take an example from the social sciences, whereas on one psychological theory, for example, Freud's, the predicates "is repressed" and "is

[24]We owe this suggestion to Donald Davidson. One aspect of it is developed in his note, "Emeroses by Other Names," *Journal of Philosophy*, LXIII (1966), 778–80.

[25]Aristotle first elaborated this concept of a physical object. For him, events, at least those of the simplest kinds, are changes in and to objects.

polymorphously perverse" are projectible, on certain other psycho logical theories they are not. The theory determines the range o projectible predicates; therefore, it determines the range of lawlik generalizations.[26] If you like, it determines to a very great exten what the *facts* are. For example, Kuhn writes that after about 163

> . . . and particularly after the appearance of Descartes' immensel influential scientific writings, most physical scientists assumed tha the universe was composed of microscopic corpuscles and that al natural phenomena could be explained in terms of corpuscula shape, size, motion, and interaction. That nest of commitment proved to be both metaphysical and methodological. As metaphysical it told scientists what sort of entities the universe did and did no contain: there was only shaped matter in motion. As methodological *it told them what ultimate laws and fundamental explanations mus be like*: laws must specify corpuscular motion and interaction, an explanation must reduce any given natural phenomenon to corpuscu lar action under these laws. More important still, *the corpuscula conception of the universe told scientists what many of their researc problems should be.*[27]

Notice that, given the resolution of the "grue paradox" we ar considering, there is no general way to distinguish between projecti ble and nonprojectible predicates. For example, one cannot tel which ones are projectible just by looking at them. Projectibilit ultimately is relative to particular theoretical contexts: predicate that are projectible in one theoretical context are not necessarily i another. In precisely the same way, there can be no general cha acterization of the notion of a positive instance. Once again, wh counts as a positive instance is relative to particular theoretica contexts.[28] This line with the "paradoxes of confirmation" seem to lead us in the direction of quite a sweeping claim: If the proble of confirmation (or induction) is to give such a theory-independen general characterization of instancehood or projectibilty, then it h no solution. The questions "What evidence confirms or support this theory?" and "Which generalizations about emeralds are law like?" are legitimate; the questions "What constitutes positive ev dence for a theory?" and "Which predicates are projectible?" a not.

[26]As was suggested in connection with the "raven paradox," the theory al determines the range of positive instances.

[27]*Op. cit.*, p. 41, italics supplied.

[28]The observation report that the sun goes around the earth is a positive i stance, and thus confirms Ptolemy's geocentric hypothesis. On the Copernica theory, the same observation report is held to be based on some sort of optic illusion!

To say that the projectibility of predicates is relative to a particular theoretical context, however, is not to say that those predicates that are projectible are determined by the primitive vocabulary of the theory. It is not the case, for example, that motion can be attributed to corpuscles simply because certain predicates are primitive in a given formulation of the corpuscular theory space-time. Similarly, "grue" can be defined in terms of what predicates we assume to be in the primitive vocabulary of a theory about emeralds, that is, "is green before *t*," "is blue after *t*," and so on. Other considerations, if any, rule out "grue" type predicates.

Suppose we define a new predicate, "Q," in terms of two primitive predicates of classical (Newtonian) physics: "Q*a*" just in case *a* has a determinate position at time *t* or *a* has a determinate velocity at time *t*, but not both. "Q" can be *defined* within classical physics—we've just done it—but it cannot be *projected* within it. Why not? Because as construed in classical physics, it is not possible that the objects studied by that particular physical theory do not have both a determinate position and velocity simultaneously. Quantum theory, in projecting "Q" at the same time reconstrues the objects that the earlier theory studies.

This last remark leads to another. Some philosophers, notably Kant, on one interpretation of his position, have argued that although theories may come and go, they do so only within a conceptual framework, which is itself inviolate for a variety of reasons. In particular, an Aristotelian type of concept of a physical object is part of that framework, and as a result under no circumstances can "grue" type predicates be projected.[29] But this line would, it seems, tell equally against certain developments hinted at in our brief remarks about quantum theory. In fact, a small group of physicists has argued against the adequacy of quantum theory on the ground that it conflicts with a particular concept of a physical object, and so eventually with particular concepts of a physical law and a physical explanation. The issues here are too complex for us to do more then mention them. The point remains that if, in certain circumstances, and some would claim these circumstances are unimaginable or unintelligible, "grue"" type predicates were ever projectible with respect to emeralds, then in those circumstances, there would have been some fundamental changes; either emeralds would no longer

[29]The conceptual framework, Kant seems to have thought, determined the shape of any possible physical theory, hence the range of projectible predicates, hence the range of positive instances, hence the range of lawlike generalizations. All that is needed, he thought, to solve the "problem of induction" is to show how and in what ways this framework is "necessary."

be regarded as typical physical objects or our concept of a physical object would have changed. The same kind of choice seems to confront us in accepting quantum theory; for example, either mesons are not physical objects or our concept of a physical object has changed. To rule out the latter possibility, as some philosophers have done on a priori grounds, is, in turn, either to reject the quantum theory—more accurately, to reject a given interpretation of its mathematical symbolism—or to deny to the objects it investigates physical-object status.

4. Are the Laws of Nature True or False?

The heavy reliance put on the notion of *theoretical context* in discussing the "paradoxes of confirmation" raises the question: if theories determine, at least in part, what is going to count as evidence for them, how can we speak of their being confirmed? After all, isn't it the job of theories to interpret or classify what we already know to be fact? Further, how is the adequacy of a theory to be assessed, if not by seeing how it holds up when confronted by the facts? In contrast, what has been said in the previous section suggests that theories determine facts—at least to the same extent that facts determine theories. If there are not theory-independent facts, then how can we say that one theory is better than another, or that our knowledge of the physical world steadily increases, or that we are nearer to the truth than we once were?

These questions point to consequences of the use of the notion of theoretical context and, indeed, are very difficult to answer. For one thing, the picture of science and knowledge that these questions reflect is deeply entrenched. Theories are tested by confronting them with facts and confirmed by those that accord with them; there is only one true or correct theory about some subject; theories are "discovered" not "invented," and so on. This picture not only is entrenched as the correct description of the relation between fact and theory, but also frequently serves as a regulative ideal for scientists, much of whose work seems to be motivated by the attempt to discover *true* laws. For another thing, the alternative picture of the relation between theory and fact suggested in some of the discussion on confirmation has not yet been elaborated in any detail, and remains somewhat programmatic.

Still, certain aspects of the alternative picture can be sketched. Perhaps of greatest importance is the implication that if theoretical context determines the form and content of our generalizations about nature, then these generalizations confirm the context, if at all, only very indirectly. The context serves rather as a system of leading or first principles that make confirmation of generalizations formulated with respect to it possible; thus it determines what is to count as evidence and lawlikeness, and how research is to be carried out. Therefore, it seems to make little sense to talk about evidence that confirms or supports these principles, or principles being confronted with the facts; indeed, some have urged, it makes little sense to say that these generalizations are either true or false!

In the alternative picture of the relation between theory and fact we are considering, the leading or first principles might be construed in analogy with moral rules that, though neither true nor false, determine the context in which behavior is both described and appraised. The rules themselves are not descriptions of anything; they are normative or regulative, but they make description and moral judgment possible. There are other respects in which the analogy with moral rules breaks down. For example, if natural laws are broken, we conclude that they are not laws, while if moral rules are broken, we do not conclude that they are not rules; but insofar as first principles have no truthvalues, are context creating —in fact presupposed by particular methods of scientific description or moral appraisal—and only implicit in science, the analogy with moral rules is unmistakable.

One problem for this view of first principles is that these same principles and rules seem arbitrary—a feature that conflicts with the intuition that some principles and rules are clearly *better* than others. The intuition is not groundless; how else can one explain the abandonment of one theoretical context for another, for example, Newtonian physics for relativity theory? However, it might be urged that this problem stems from two sorts of mistakes. The first is the confused idea that because first principles are *conventional*, they are *arbitrary*. To say that they are conventional is to say that we have some choice in the matter of such principles. Nature does not dictate a unique choice; decision on our part is required. In other words, if theories are, in T. S. Kuhn's words, "ways of seeing the world," there is more than one of them. The second point of confusion is that because first principles are conventional, there is, therefore, no means of comparing them, no basis on which to reject some in favor of others. To be sure, we cannot compare two theories by confronting them directly with the facts, as Descartes and Bacon suggest in their emphasis on the importance of *crucial experiments*; for two genuinely different theories—for example, Copernicus' and Ptolemy's—determine different facts. However, this does not mean that they cannot be compared. Comparisons can be made on the basis of internal coherence, elegance, simplicity (with all the difficulties in the notion) richness of deductive consequences, and so on. Since all these enter into a consideration of the adequacy of theories, it is certainly not an arbitrary matter which theories we adopt. Even if we give up the idea that some theories are *true* and others false, we retain the idea that some are more *adequate* than others. On the other hand, it is impossible to say in advance which of these factors is the most important or deciding factor; there are no acceptance or rejection procedures of standard sorts for theories. The reasons some theories are rejected in favor of others are not always the same.[30]

These remarks do not apply to the numerous generalizations that are formulated and tested within a theoretical context. There are rejection and acceptance procedures of standard sorts for these generalizations, as established by that context. Typically the acceptance procedures involve confirmation via positive instances.[31] However, the problems that arise in connection with them have more to do with statistics and probability theory, and are as much a problem for the mathematician as they are for the philosopher.

Most of the scientific enterprise seems to take place within well-defined theoretical contexts. It is for this reason perhaps that philosophers intent on describing that enterprise have tended to ignore or downplay the context, on the one hand, and to stress the centrality of well-confirmed empirical generalizations, on the other. The alternative picture of the relation between fact and theory we have been describing suggests, however, that the more traditional picture of the relation between fact and theory has created for itself

[30]It certainly is not the case that theories are accepted over others if they answer the same questions better. Typically, they ask and answer different questions, for example, when anti-Copernicans asked why, on the heliocentric theory, objects didn't fly off the earth, the Copernicans had no answer. The reason they didn't immediately abandon their own theory was their belief that other questions, which they could answer, were very much more relevant. Another, more contemporary illustration of the point, is the following. Sometimes psychologists in the Skinnerian tradition are criticized for over concentration on the shaping or molding of behavior. For example, to the question, "Why does the rat press the bar *before* reinforcement can take place?" there usually is no answer given. Indeed, there may be no answer to this question (it might be judged meaningless in Skinnerian theory), or the answer, if there is one, awaits elaboration of the theory. Other psychological theories of learning, for example, Tolmanian expectancy theory, regard an answer to this question as essential to determining the acceptability of a theory of learning.

[31]This claim must be taken with a grain of salt. Often generalizations are accepted because they are derived from other, well-founded generalizations.

the problems of confirmation. Emphasis on context serves to dispel such problems, though, to be sure, it leaves us with nagging questions about the truth of the laws of nature.

For Further Reading

Carnap, Rudolf, *The Logical Foundations of Probability*. 2nd ed. Chicago: University of Chicago Press. 1962. This is an advanced text. We mention it here only because the work is especially exhaustive and authoritative.

Goodman, Nelson, *Fact, Fiction, and Forecast*. Indianapolis, Indiana: The Bobbs-Merrill Company, Inc., 1965.

Kuhn, T. S., *The Structure of Scientific Revolutions*. Chicago: University of Chicago Press, 1962.

Salmon, Wesley, *Foundations of Scientific Inference*. University of Pittsburgh Press, 1967.

Scheffler, Israel, *The Anatomy of Inquiry*. New York: Alfred A. Knopf, Inc., 1963.

Skyrms, Brian, *Choice and Chance*. Belmont, California: Dickenson Publishing Co., Inc., 1966.

THE LIMITS OF SCIENTIFIC EXPLANATION

5

1. Introduction

In Chapter 1, we suggested that any attempt to determine the limits of scientific explanation depends on how the concept of scientific explanation is analyzed. This point is so trivial that it would not be worth making, except for the fact that many people at various times have claimed that "science," where this term is left vague and unanalyzed, has certain intrinsic limits, period. There are facts that "science" per se cannot explain. But suppose, for example, we accept the "Humean" account, that scientific explanation essentially includes reference to laws. It follows that we cannot explain or predict events under *every* description of them. For any two distinct events cannot be completely described in the same way (that is, their being *distinct* events ensures that there is at least one property they do not share). Further, it seems reasonable to require of any statement that to be a law it must subsume or "cover" at least two distinct events.[1] Hence an event cannot be subsumed or "covered" by a law under every description of the event. Explanation and prediction on the "Humean" account are always with respect to some, but not all, aspects of events. They are always "limited" in this way. Much the same point was made earlier in discussing LaPlace's claim that given the positions and momenta

[1]To use a "law" with but a single instance to explain that instance would be patently circular.

of all objects in the world, and Newton's Laws, every future state of the world could be predicted. The truth is that given such information the future could not be predicted in full detail; our description of events would be restricted very roughly to spatio-temporal properties of them.[2] On the other hand, if we don't accept the "Humean" account of explanation, and maintain instead (as some philosophers have) that to explain scientifically is to describe in the greatest possible detail,[3] then the limits of such explanation will not lie in this direction. They will, for example, lie in the direction of not being able to predict future states of affairs in a lawlike way.

Even given a particular concept of scientific explanation, further restrictions on our discussion are desirable. In particular, we want to avoid two possible extremes. Once again, given the "Humean" account, one extreme is to argue that since this account essentially includes reference to laws, and since such laws rest on inductive inferences for which there is no justification, science explains nothing. It takes place in a kind of epistemological vacuum. This is the way of traditional scepticism: to undermine the pretensions of the scientific enterprise at its foundation. But this sort of argument is not our concern here. The question is not whether science can explain anything, but whether there are certain things, for example, certain types of phenomena, that it cannot explain.

The other extreme is to argue that scientific explanation has no limits. This position has two variations. One is to claim not that scientific explanation has no limits, but that whether or not it does is an empirical question. Such limits cannot be determined on conceptual or "philosophical" grounds, a priori. It is difficult to see how a general argument in support of this claim would go. Our interest, in any case, will be in seeing whether and where the particular a priori arguments we consider succeed; their failure does not imply that the question is not conceptual. The other variation on this extreme position is to claim that the development of science has been marked by the extension of scientific explanation to ever new and broader kinds of phenomena. On quasi-inductive grounds, we have reason to expect that eventually everything will be capable of

[2]For example, we could predict in what direction this billiard ball would move, and with what velocity, and so on, but we could not predict color changes it might undergo, or future ownership, and the like.

[3]The claim might run as follows: "The more we know about the event the better we understand it. God, who alone can give complete descriptions of events, has perfect understanding of them."

being scientifically explained. But the question is not whether science can "explain everything" (whatever that means), but whether there are, for example, observable phenomena, which it purports to, but cannot (for conceptual reasons) explain. In the case of the "Humean" account, this comes to asking if and/or why there are events that are not "covered" or cannot be "covered" by laws. We are not, for example, interested in whether there are certain ultimate questions that only religion can answer.

Within the restrictions on the discussion thus proposed, we will take up three different arguments that attempt to determine the limits of scientific explanation.

2. Determinism and Explanation

The first argument is to the effect that human actions, a rather generous class of observable and otherwise unmysterious events, cannot be scientifically explained. For if they could be scientifically explained it would follow that they were determined. But if they were determined, it would follow that they were not voluntary. But human actions, or at least a great many of them, are voluntary. Therefore, human actions cannot be scientifically explained. Here is one place where limits must be drawn.

As we have stated it, the argument has three premises. The first asserts a connection between explanation and determinism, the second a connection between being determined and being involuntary, and the third asserts the (presumed) fact that at least some actions are voluntary. Often the argument is given an ethical twist: The first two premises are supplemented by a third premise—if actions were not voluntary, it would follow that we could not be held morally responsible for our actions. We will be mainly concerned with the original statement of the argument, however.

The conclusion seems to follow from the premises. The choice, therefore, is either to accept it or question them.

Perhaps the most straightforward way to avoid the conclusion is simply to reject the third premise, that at least a great many human actions are voluntary. That many human actions are voluntary, it is contended, is an illusion; the impression many of us have that we often choose freely to perform certain actions is mistaken. Typically, the "illusion" is traced to three different sources; we think we act freely only because we are ignorant of the determinants of our behavior, because our use of moral vocabulary depends on it, because to think otherwise is psychologically stultifying. With a more developed theory of human action, behavioral or neurophysiological, and a consequent understanding of its real origins, we would realize that no actions are voluntary. The maze in which we live is on a much larger scale and vastly more complicated, as is the system of positive and negative reinforcements, but it is in principle not unlike that through which we run rats. Our "choices" are involuntary responses to stimuli. Some psychologists believe this, and accept as its corollary that the moral vocabulary should be scrapped, for example, one should no longer call actions "good" or "bad," or hold men morally responsible for their deeds.

There are two unhelpful replies to this claim. One reply is we "just know" (perhaps on the basis of introspection) that we choose freely; one is "aware" at the time of making a decision that it is voluntary, and so on. This reply is unhelpful because it doesn't take us very far. In particular, it doesn't show why our "awareness" might not be mistaken. The other unhelpful reply is to say that the claim that human actions are not voluntary is paradoxical or self-defeating—the claim itself must have been made within the maze, as by the rat. This reply is unhelpful because it does not follow from the fact that the claim is self-referential in the way indicated (that is, because to claim that no actions are voluntary is to act) that it is *false*.

A more helpful reply is to direct our attention more closely to the words "voluntary," "involuntary," "determined," responsible," "excusable," "accidental," and so on, to see what the criteria are on the basis of which they are used and whether such use is incompatible with the development of psychological theories that allow for the prediction of human behavior. To do this is to direct our attention more closely to the first two premises of the argument. In fact, claims that no human actions are voluntary are often motivated by the beliefs that these premises cannot be questioned and that a science of human action is possible. To the extent that these premises are undermined, of course, the necessity for making the prima facie extremely implausible[4] claim that no human actions are voluntary diminishes.

The first premise asserts a connection between explanation and determinism. But what sort of connection is not initially clear. The thesis of determinism, as we noted in the section on explanation and prediction in Chapter 3, can be understood in a variety of ways. Two are particularly relevant to our present discussion. One, the "weak' thesis, stems from the "Humean" account of explanation. On that account, to explain an event is to "cover" it with a law

[4]Some would say "incoherent."

(or laws). In the sense that given the laws, and statements describing certain antecedent conditions, one could have predicted the event,[5] one might want to say it was "determined." The other, "strong," thesis of determinism is that the covering laws themselves are deterministic, that is, they allow for the *retrodiction* as well as the prediction of events, as typified in classical physics and some S-R learning theories like that of Hull.

These are separate theses. But they are often jumbled together. Thus it is frequently asserted that since determinism has broken down in physics with the advent of the statistical laws of quantum mechanics, and perhaps also in psychology with respect to S-R learning theories, the "Humean" account will have to go. But this is a non sequitur. The laws that the "Humean" account requires can be either deterministic or statistical. As we interpreted the "Humean" account, to insist that scientific explanations must contain deterministic laws is to go far beyond it. Hence, it does not follow from the breakdown of determinism in the "strong" sense that certain kinds of events cannot be explained scientifically, that is, à la Hume.

This non sequitur is usually compounded with another non sequitur. Many people, including some respected scientists, have argued that Heisenberg's "Uncertainty Principle" proves that human actions are voluntary. But of course this astonishing result, however welcome, does not follow. What follows is only that one cannot argue for the claim that no actions are voluntary on the basis of the alleged fact that the laws of physics are deterministic. In particular, even with the "Uncertainty Principle," predictions on the basis of quantum theory are still possible. What has to be shown in addition is that the type of predictions that it allows is not incompatible with the notion of a voluntary action.[6]

In fact, the introduction of statistical laws does effect the type of predictions made and hence the sense in which the "Humean" account is deterministic. As we pointed out earlier, the introduction of statistical laws forces us to allow for the possibility that the connection between *explanans* and *explanandum* in an explanation is inductive, rather than deductive. But when the connection is inductive, the *explanans* does not exclude the possibility that the event to be explained did not occur. Or, symmetrically, when the connec-

[5]A statement describing it follows logically from the law(s) and statements of the antecedent conditions.

[6]Needless to say, this does not suffice to *prove* that human actions are voluntary.

tion is inductive, the *explanans* does not exclude the possibility that the event predicted will not occur. Its nonoccurrence is compatible with the explanation given.

This fact can be elaborated. It is often maintained that an action is free if the agent *could have done otherwise*; that is, this is what it means to say that an action is free. But, we have just seen, explanations and predictions of events, and hence of actions, by way of statistical laws do not exclude the possibility of particular events not occurring. Which is to say, on at least one sense of "could," that the prediction or explanation of an event is not incompatible with claiming that the agent could have done otherwise. What this comes to, it should now be clear, is the claim that in this sense "voluntary" actions are compatible with their being inductively explained or predicted, that is, with the "weak" sense in which scientific explanation à la Hume is deterministic.

One difficulty with this line of argument is that it depends on rather far-reaching assumptions about the form of laws of nature, especially those, assuming for the moment that there are such laws, governing human behavior. We have no right to assume, it might be argued, that such laws will inevitably be indeterministic, any more than we have a right to assume they will be deterministic. The freedom/determinism question can be settled only by showing either that on any sense of the word human actions are not "determined"—hence not even in the "weak" sense implied by the "Humean" account of scientific explanation, or that even the "strong" determinist thesis is compatible with the notion of voluntary action.

This latter tack is the one taken most frequently with regard to the second premise of our argument. Even "strong" determinism, even prediction, in a deductive way, in principle of every human action does not entail, it is argued, that no human actions are voluntary.

Two different sorts of arguments to show that determinism, however construed, and freedom are not incompatible are classic. The first derives from Aristotle.[7] An action is voluntary. Aristotle seems to suggest, not when it is uncaused, but when the cause is of a particular sort or, as he says, "lies within." The appropriate "internal" causes are beliefs, desires, reasons, motives, and so on, and not, for example, muscle spasms. When the action in question is caused by items of this kind, it can be said to be voluntary.

[7]What follows is a vastly oversimplified account of the argument in the third and sixth books of the *Nichomachean Ethics*.

This way of reconciling freedom and determinism has much to recommend it. For one thing, it connects up the notions of an action's being voluntary and an agent's wanting to do something in a very intuitive way. For another thing, it rules out, again very intuitively, all those actions that are caused by various "external" circumstances, coercion, hypnosis, and so on. For a third, it makes sense of the fact that we have control over many of our own actions. Too often those who argue for the voluntary nature of human actions suggest that such actions are completely spontaneous, uncaused, even irrational. But the notion of a voluntary action seems to include the fact that voluntary actions are under our control, that is, are caused by us in certain ways. The operative causes of actions "under our control" are desires, beliefs, and the like.

The main difficulty with the Aristotelian account for us is that it is at least controversial whether an explanation in terms of desires, beliefs, and so on, can be put in "Humean" form, that is, whether *teleological* explanations can be reduced to causal explanations.[8] Thus, one might argue that although there is nothing mysterious or inexplicable about human action, it cannot be explained in the manner most appropriate to science. In particular, it might be claimed, for reasons given in the section on "teleological explanation" in Chapter 3, that it is impossible to formulate *laws* in terms of desires, beliefs, and so on. Similarly, it might be claimed that given information about a person's motives, one cannot predict what course of action he will undertake. Indeed, it is for this reason that an action's being explained teleologically is compatible with its being voluntary. To put the issue in a nutshell: whether or not the Aristotelian account serves to reconcile freedom and determinism in the way indicated depends on whether or not teleological explanations are irreducible.[9]

The other classic argument to show the compatibility of freedom and determinism derives from Kant,[10] and is very much related

[8]Of course, if one counted teleological explanations as scientific explanations there would be no further problem. For in that case, there would be no incompatibility between explaining an action teleologically—hence in terms of a person's motives, hence as voluntary—and explaining an action scientifically. Scientific explanation would not be limited in this direction. This move might come as a disappointment to those who want to contend that there is such a thing as "freedom in an absolute sense."

[9]We should add that there are other difficulties with the Arisotelian account, of voluntary action, but they cannot be taken up here.

[10]Once again, vast oversimplification must be admitted. Our interpretation of

to the issues raised in the last paragraph. For Kant (for whom, incidentally, determining the limits of scientific explanation was the principal task of philosophy) the problem was to reconcile two fundamental principles: "Ought implies can," that is, ascription of moral vocabulary entails that the action appraised is voluntary, which he took to mean "uncaused," and "every event has a cause." Since he took actions to be events, the principles are at least prima facie incompatible.

Roughly, his solution was to say that actions are describable in two different ways—as events in the physical world to which the vocabulary of science, in particular the category of causality, is appropriate, and as events in what might be called a "mental" or "intentional" world to which the moral vocabulary is appropriate. There are not two different sorts of events, just two different sorts of descriptions. Under their physical descriptions, events can be explained and predicted à la Hume. Under their "mental or intentional" descriptions, they cannot. For under these descriptions, and for reasons already given, the formulation of the required laws is not possible.

To make the connection with our discussion of Aristotle's argument more explicit, we might say that Kant is contending that of a certain class of events, human actions, both teleological and causal, that is, "Humean," explanations can be given. Teleological explanations cannot be reduced to causal ones; they differ also, it is suggested, in that they are offered in a context in part created by the institutions of moral praise and blame. But this does not entail that certain events cannot be scientifically explained. It entails only that they can be scientifically explained only under certain of their descriptions. Conversely, that an action is scientifically explainable does not entail that it is not voluntary, even if we add that the laws that figure in scientific explanations must be of a "strong" deterministic type. Whether an action is voluntary or not depends rather on whether or not it is appropriately described in the "mental" or "intentional" vocabulary also, whether it can be given a teleological explanation.

An illustration should help. We describe one and the same event as *a man raising his arm* and *a man's arm going up*. The

---

the Kantian position owes much to Donald Davidson. See his paper "Mental Events," in *Experience and Theory*, ed. Lawrence Foster (Amherst, Massachusetts: University of Massachusetts Press, 1970). The relevant texts are *The Foundations of the Metaphysics of Morals* and the section on the third antimony in the *Critique of Pure Reason*.

first is the "intentional," the second, the physical description. Under the second description, we can give a causal explanation, perhaps in terms of the muscles contracting, and so on. Under the first, we cannot. But if the first is also an appropriate description, for example, if the man's arm did not jerk up as a result of his having been stimulated cortically, then we can say that the action is voluntary.

Kant's view is not without problems. There are, for example, difficulties in the notion of describing the same event in two different ways, that is, in giving identity conditions for events, and in saying what constitutes an appropriate description. Having mentioned them, we will say no more about them. But there is one problem connected with his view that comes up again in the last section of this chapter. It is that insofar as science explains events only under certain descriptions of them, it leaves something out. What it leaves out in the argument just sketched is their "mental" or "intentional" aspects. It is, then, not so much that there are events that resist scientific explanation, but that certain properties under a given description, or in a given framework, of these events do. Most importantly, "mental" or "intentional" properties of them do. It is with respect to these properties that scientific explanation is limited.

3. Reduction and Explanation

The second argument on the limits of science is to the effect that certain phenomena are "emergent" with respect to others and so cannot be scientifically explained. For if they could be scientifically explained, it would follow that the former phenomena could be "reduced" to the latter. In particular, it would follow that biology could be "reduced" to physics, mind to matter, life to inanimate nature. But "emergent" phenomena cannot be reduced. Therefore, many of these phenomena cannot be scientifically explained.

Despite the fact that it has a long history, and is usually advanced with a great deal of passion, this line of argument is far from clear. What does it mean to say that certain phenomena are "emergent" with respect to others? What does it mean to say that certain phenomena cannot be "reduced" to others?

We can simplify matters by defining "emergent" in terms of "reduced." Thus, those phenomena are "emergent" with respect to others that cannot be "reduced" to them. There seem to be at least two ways in which "reduced" itself can be understood. On the one hand, to reduce certain phenomena to others is to show that the laws that describe the former follow logically from the laws that describe the latter. We mentioned an example earlier: the phenomena of heat reduce to mechanical phenomena, insofar as the thermodynamical laws that describe the former can be derived from the statistical mechanical laws that describe the latter. On this account, the connection between explanation à la Hume and reduction is clear. Reduction and explanation have precisely the same logical structure.

The other way to understand "reduced" was indicated in the section on reductive explanation in Chapter 3. On this way, to reduce certain phenomena to others is to decompose the former in terms of the latter. In this sense, we reduce, or attempt to reduce, biology to physics by showing that cells are composed and can be understood in terms of, for example, atoms. The ostensible reduction of thermodynamics to statistical mechanics and the kinetic theory of matter provides another illustration, macroscopic thermodynamical phenomena are decomposed and understood in terms of microscopic mechanical phenomena.

The same sort of picture is associated with both concepts of reduction. It is that of a hierarchy of levels, with each of which there is associated a unique range of objects and properties.[11] We sketched the picture briefly in Chapter 3 in discussing Descartes' philosophy of science. Much of the controversy surrounding the issue of reduction, in fact, has Descartes' version of the hierarchical picture in the background. Thus, the question is typically not whether, for example, biology can be reduced (in either sense) to some other subject, but whether it can be reduced to *physics*. Conversely, it is maintained that unless biology can be reduced to physics, then biological phenomena have not been scientifically explained. Not only is physics the touchstone, but a particular nineteenth century mechanistic-deterministic physics. We are asked whether animate nature, specifically man, is a machine, whether a "mechanistic" position can account for "vitalistic" phenomena, and so on. But these are embellishments. The central issue is not "mechanism versus vitalism," but whether certain phenomena can be reduced, in the intended sense, to others.

[11]This is not quite right. Not all reductions of the "derivation" type are to component parts with different properties. Thus, Galileo's laws as derived from Newton's "reduce" to them, as eventually, it is usually contended, do Newton's to Einstein's. There is no shift from one theory to the other as regards the objects studied. We will ignore shift-less reductions of this sort, although they are not unproblematic.

Reduction is, then, a relation between "levels" (as crudely characterized). Understood in one way, it is a logical relation: we reduce one level to another when we derive the laws governing the former from the latter. Understood in another way, it is somewhat on the model of the part/whole relation. Similarly, "emergence" is a relation between "levels." Phenomena are emergent, in the first sense, with respect to phenomena on another level, when statements describing the former cannot be derived from statements describing the latter. Phenomena are emergent, in the second sense, when they cannot be decomposed in the way indicated.

The so-called doctrine of "emergence" claims that certain phenomena cannot be reduced to others, hence are emergent with respect to them, hence cannot be scientifically explained. It has two variations. The first claims merely that in the theories that *presently* characterize the different "levels," reduction, in either sense, is not possible. Given the present state of physics, for example, a reduction of biology is out of the question. What is "emergent" today might be reducible tomorrow. Opponents of the doctrine of "emergence" like to say that it turns on what state science is in, a contingent matter.[12] In fact, its proponents, for example, Spencer, Alexander, Morgan, and Broad, claim that the reducibility of certain types of phenomena is impossible in principle.

C. D. Broad, for instance, argues in the following way: Macroscopic qualities of objects cannot be reduced to and explained in terms of other properties of their parts, because this would entail that, given the parts and their properties, other macroscopic qualities could be predicted in advance of their having been observed, and this is impossible. For example, he asserts that we could not "possibly have formed the concept of such a color as blue or such a shade as skyblue unless we had perceived positive instances of it." As a result, even a LaPlacean superman could not deduce from his knowledge of the microscopic structure of a given object that the object would appear blue or sky-blue as it does to human beings. "If the existence of the so-called 'secondary' qualities," he continues,

> or the fact of their appearance depends on the microscopic movements and arrangements of material particles which do not have

[12]For example, Ernest Nagel dismisses the pretended implications of the "emergentist" position with the remark that it has failed to observe that "novelty is a relational characteristic of properties with respect to a given theory." In A. Danto and S. Morgenbesser, eds., *Philosophy of Science* (New York: World Publishing Company, Meridian Books, 1960), p. 311.

these qualities themselves, then the laws of this dependence are of the emergent type.[13]

Many different points are mixed into the argument, including a particular theory of concept formation. Three points are crucial. The first is that Broad assumes the symmetry of explanation and prediction. Failure to explain is failure to reduce is failure to predict. The second point is that Broad's argument seems to rest on what many have contended is a simple fact of logic:[14] No term that does not occur in the premises of a valid argument can occur in the conclusion. From premises that describe the movements and arrangements of material particles, we cannot derive conclusions about their colors. Or, to take another example, from premises (for example, the fundamental assumptions of statistical mechanics) that do not mention "temperature," we cannot derive conclusions (for example, the laws of classical thermodynamics), that do. The third crucial point is that Broad runs together the two concepts of reduction. He wants to maintain that there are certain things that mechanical-atomistic theories cannot explain. His reason is that macroscopic qualities of objects cannot be reduced in the sense that statements describing them cannot be derived from these theories. Reduction in the decomposition sense is impossible because reduction in the derivation sense is.

Assume that the logical point must be granted: no term that does not occur in the premises of a valid argument can occur in the conclusion. Then, if phenomena on one "level" are to be reduced to (derived from) phenomena on another "level," there must be a way of connecting the concepts that occur on one level to concepts that occur on the other. Two suggestions as to the nature of this connection have received most attention. One is that the connection is "empirical." By this is meant simply that in fact the concepts on different levels that we wish to connect refer to the same state of affairs. "The Morning Star" is not synonymous with "The Evening Star." But the two expressions refer to the same object, the planet Venus. Further, it took many years of astronomical observation to determine that this was the case. Similarly, the connection between "temperature" and "mean kinetic energy" that allows thermodynamics to be reduced to mechanics is that the two expressions refer to the same state of affairs. The identity between them is

[13] C. D. Broad, *The Mind and Its Place in Nature* (London: Routledge & Kegan Paul, Ltd., 1925) pp. 71–72.

[14] Some philosophers have regarded this as controversial. Cf. Nagel, *The Structure of Science*, p. 353n.

not one of meaning but of fact. Reduction, so those who make this move claim, requires the introduction of such factual identities.

The difficulty with this view is that in the course of reduction the meanings of the terms only contingently identified seems to change as well. Once "temperature" is identified with "mean kinetic energy," for example in the reduction of thermodynamics to mechanics, it is applicable to a variety of circumstances and with a meaning that it was not previously.[15] This fact gives rise to a second suggestion, that in the course of reduction, concepts on one "level" are *redefined* in terms of those of another "level." The connection between them is thus quasi-conceptual. This view has much to recommend it. The difficulty is that if, for example, the concepts of thermodynamics or biology are redefined, then in what sense is it that thermodynamics or biology as we originally understood them have been *reduced*. For the "emergentist" wants to claim that, given the present meanings of the terms in those sciences, they cannot be reduced; the redefinition suggestion admits as much.

Still, it is possible to avoid Broad's conclusion, if we first distinguish between the two concepts of reduction and then maintain that atomistic-mechanical theories are reductive, and explanatory, in the decomposition sense only. The suggestion is that the two concepts are incompatible: attempts to yoke them together in a single account inevitably lead to difficulties, for example, the "emergentist" position. If the derivation sense of reduction, and perhaps also of explanation, that is, the "Humean" account, is abandoned, a Broad-type argument cannot be used to prove the existence of "emergent" properties and the fact that science implies its own limits in this direction.

There is another type of argument, however, which is directed more specifically against reduction in the decomposition sense. Suppose we assert that water is composed of hydrogen and oxygen, hence, in the intended sense, reducible to them; its properties can be explained in terms of the properties of the chemical elements. Or, to take a more problematic case, suppose we assert that sensations are made up of events in the central nervous system, hence reducible to them; the phenomena of sensations are explained by

[15]Only if we allow that "temperature" changes meaning in the course of reduction can we claim that thermodynamics can be derived from mechanics and the kinetic theory of matter, for the classical theory of thermodynamics is otherwise inconsistent with mechanics and the kinetic theory. Cf. Karl R. Popper, "Irreversibility: or, Entropy since 1905," *British Journal for the Philosophy of Science*, VIII (1956–58), 151–55.

neurophysiology. The argument is that water or sensations, or more accurately their properties (for example, "is a thirst-quencher," "is blue") are "emergent" with respect to oxygen and hydrogen, on the one hand, and events in the central nervous system, on the other. For to say that phenomena can be reduced to other phenomena in the sense of being decomposed in terms of them is to say that the former can be identified with the latter. Thus, a pail of water on this concept of reduction is identical with a given collection of hydrogen and oxygen atoms, a sensation is identical with an event or events in the central nervous system. But this is impossible. For if, for example, a pail of water could be identified with a collection of chemical elements, then it should be possible to attribute the same properties to each. But this cannot be done. Being a thirst quencher is a property of water, having a numerically determinate atomic weight is not. Conversely, being a thirst quencher is not a property of chemical elements, although having an atomic weight is. Hence, water and combinations of chemical elements cannot be identified. Hence, the former cannot be reduced to the latter, and so on.

The water-chemical element identification has rarely been questioned. But the same pattern of argument is often used when one tries to reject a reductive account of psychological phenomena, or perhaps more prominently, the so-called "secondary qualities." The "secondary qualities" (traditionally: color, sound, taste, smell, heat, and cold) cannot be reduced to the "primary qualities" (traditionally: size, shape, and motion), because it would follow that the "secondary qualities" could be identified with the "primary qualities." And this is not possible, for it would mean that both sets of properties had the same properties. But they do not. For example, "secondary qualities" are nonspatial, "primary qualities" are spatial. In the same way, it is urged that sensations cannot be identified with events in the central nervous system because, for example, while sensations are blue, neurophysiological events cannot be.

This argument has always found a great deal of favor, even with those who hold no brief for the doctrine of "emergence." But there are replies to it. We will mention three. The first two echo replies to the Broad-type argument considered a moment ago. One contends that the identity between, for example, a pail of water and a combination of chemical elements is of a factual or "empirical" or contingent kind. To insist that if two objects are to be identified that they must have all and only the same properties in common is to insist that the identity between them is of a "stronger," conceptual kind. The second reply contends that in the sense in which reduction leads to identification, the concept of water has been redefined. It *now* makes sense to say that water has esoteric chemical properties, although it did not formerly. Just as with the development of the corpuscular theory in the seventeenth century, one might want to claim that the concept of a physical object changed, so that its properties became essentially those of the corpuscles of which they were allegedly composed and with which identified. For example, on the corpuscular theory, color could no longer be attributed to corpuscles; so too, color could no longer be attributed to large-scale physical objects, but was instead a property of perceiver-dependent sensations.

We have already said where the difficulties in the first of these replies lie. They revolve around the notion of a factual or "empirical" connection between concepts. Perhaps they are not overwhelming. The third reply is to say that it is a mistake to think that reduction implies identification. What happens, rather, is that we *replace* our old concept of water and talk about water with the concept of chemical elements and talk about chemical elements.[16] Reduction is replacement, more strongly, elimination, not identification. The properties and objects held to be "emergent" are no longer of any account in physical theory. Generalizations about water are eliminated in favor of generalizations about chemical elements. Or, to use another of our examples, in the course of the shift to the corpuscular theory in the seventeenth century, color became not an "emergent" property of objects, but an *irrelevant* property of them.[17] But this reply in turn gives rise to a new argument that attempts to put limits on scientific explanation. It states that since science ignores certain properties or phenomena at the expense of others, something, for example, the "secondary qualities," is "left out."

4. Description and Explanation

The third argument on the limits of scientific explanation is that scientific explanation is limited by the fact that it is always with respect to certain properties or phenomena. Other properties and phenomena are ignored. Further, since these

[16]Ordinary language does not always reflect the shift. We still talk (and make perfectly good sense) about the sun setting in the west, stones being impenetrable, and so on.

[17]It might be argued that sensations are "disappearing" in a precisely parallel way.

latter are not to be ignored, that is, they form an essential part of any adequate conception of nature, scientific explanation can never be more than partially satisfactory—it can never lead to genuine understanding of the natural world.

The flavor of this argument is best caught in examples. The explanations that classical physics offers of the behavior of physical objects, for one, is with respect to only certain properties, for example, size, shape, mass, and motion of those objects. Other, "secondary," properties of them are ignored. They are "left out" of the scientific picture of the world.[18] As a consequence, that picture is limited. Perhaps for the same reasons, it is inadequate.

Or, to take another example, it is often asserted that a hallmark of scientific explanations is that the properties in which they deal are *quantitative* or measurable. But this is to "leave out" the *qualitative* properties of things.

Or, to take still a third example, it is often contended that the properties in which scientific explanations deal are publicly observable, those which occupy positions in space and time. But this is to leave out other properties, notably mental phenomena, which are not spatio-temporal.[19]

It is traditional, following Descartes, to run these three examples together. Thus, the "secondary qualities" are held to be both qualitative and mental. More generally, it is held that science bifurcates nature. On the one side are physical objects and properties, the quantitative, invariant, spatial, objective phenomena; on the other side are, very broadly, "mental" objects and properties, the qualitative, variant, nonspatial, subjective phenomena. But insofar as it bifurcates nature in this way, scientific explanation is inadequate. The conception of nature it offers us is distorted, one-sided.

The first comment to make on this argument is that it depends on augmenting our original concept of scientific explanation in a number of ways, in particular by requiring that the laws on which such explanation rests deal with the quantitative, spatial,

[18]Cf. Alfred N. Whitehead, *Science and the Modern World* (New York: Mentor Books, 1948), p. 55. "Nature is a dull affair, soundless, scentless, colorless; merely the hurrying of material, endlessly, meaninglessly. However you disguise it, this is the practical outcome of the characteristic scientific philosophy which closed the seventeenth century."

[19]Cf. Brand Blanshard, "Fact, Science, and Value," in *Science and Man*, ed. R. N. Ashen (New York: Harcourt, Brace & World, Inc., 1942). "With the faintest and simplest element of consciousness, natural science meets something for which it has no pigeon-hole in its system . . . Mind at its best is autonomous."

nonmental, and so on, properties of things. To put it a slightly different way, nothing in our original analysis entails that science must "leave out" the "secondary qualities," mental phenomena, and so on.

The argument is actually not directed against the "Humean" concept of explanation. Rather, it attacks a particular development of science, by and large that of classical physics. But we are not interested in the explanatory limits of particular theories. Our concern is with the question of whether there are a priori reasons that restrict a general pattern of explanation to certain kinds of events and properties to the exclusion of others. And there is nothing in the "Humean" account, for example, that demands that an explanation be in terms of the measurable properties of things.

On the other hand, there is something to be said for a more general form of the argument, one that is not dependent on any given stage of the scientific enterprise. For it does seem to be the case that scientific explanation inevitably involves a bifurcation of nature, although not necessarily at the same point as classical physics. One reason has to do with the logical character of laws. As we indicated in the first section of this chapter, laws cannot completely describe events and at the same time "cover" two or more of them. They must always be in terms of only certain aspects. Certain other aspects will, as a result, be "left out." A second reason why scientific explanation inevitably bifurcates was mentioned earlier, in the section on "reductive explanation" in Chapter 3. We there quoted the physicist Heisenberg to the effect that if a theory is genuinely explanatory of certain properties of objects, then it cannot itself be formulated in terms of those same properties. To use our earlier example: To say that a table has a given color because it is composed of parts that have that color is to postpone rather than to provide an explanation. A cursory glance at the history of science underscores the claim: most if not all explanatory theories have been in terms of properties other than those that they wished to explain. Thus, there is an inevitable distinction between "primary" and "secondary" qualities although, once again, not necessarily where classical physics draws it.[20]

Perhaps we can say, then, that although scientific explanation inevitably bifurcates nature, it is not because such explanation is restricted to a set of properties determined in advance. There is

[20]In fact, contemporary physics uses a different set of properties, for example, spin, charge, in terms of which these properties formerly thought of as "primary" can themselves be explained.

always a distinction to be made between "explained" and "explaining" properties. But which properties are which will depend on the theories at hand and the purposes to which they are put. Hence we cannot draw, from this direction, any limits as to the type of phenomena that can be scientifically explained. Nothing in the concept of scientific explanation precludes any property from serving either as explanatory or explained. This is, of course, to characterize the scientific enterprise very broadly. But it is a very broad enterprise.

For Further Reading

Broad, C. D., *The Mind and Its Place in Nature*. London: Routledge & Kegan Paul, Ltd., 1925.

Nagel, Ernest, *Sovereign Reason*. New York: The Free Press, 1954.

Sellars, W. S.,"Philosophy and the Scientific Image of Man," in *Science, Perception, and Reality*. New York: Humanities Press, Inc., 1963.

Whitehead, Alfred N., *Science and the Modern World*. New York: Mentor Books, 1948.

INDEX